Thinking in Physics

Strategies for Improving Scientific Reasoning,
Conceptual Understanding, and
Problem Solving in Introductory Physics

Vincent P. Coletta
Loyola Marymount University, Los Angeles

PEARSON

Boston Columbus Indianapolis New York San Francisco Hoboken
Amsterdam Cape Town Dubai London Madrid Milan Munich Paris Montréal Toronto
Delhi Mexico City São Paulo Sydney Hong Kong Seoul Singapore Taipei Tokyo

Publisher: Jim Smith
Executive Editor: Becky Ruden
Project Manager: Katie Conley
Editorial Assistant: Sarah Kaubisch
Cover Designer: Derek Bacchus
Manufacturing Buyer: Stacey Weinberger
Marketing Manager: Will Moore
Cover Photo Credit: iStockphoto/360/Getty Images

Credits and acknowledgments borrowed from other sources and reproduced, with permission, in this textbook appear on the appropriate page within the text.

Microsoft® and Windows® are registered trademarks of the Microsoft Corporation in the U.S.A. and other countries. Screen shots and icons reprinted with permission from the Microsoft Corporation. This book is not sponsored or endorsed by or affiliated with the Microsoft Corporation.

Copyright © 2015 Pearson Education, Inc. All rights reserved. Manufactured in the United States of America. This publication is protected by Copyright, and permission should be obtained from the publisher prior to any prohibited reproduction, storage in a retrieval system, or transmission in any form or by any means, electronic, mechanical, photocopying, recording, or likewise. To obtain permission(s) to use material from this work, please submit a written request to Pearson Education, Inc., Permissions Department, 221 River Street, Hoboken, New Jersey 07030. For information regarding permissions, call (847) 486-2635.

Many of the designations used by manufacturers and sellers to distinguish their products are claimed as trademarks. Where those designations appear in this book, and the publisher was aware of a trademark claim, the designations have been printed in initial caps or all caps.

ISBN 10: 0-13-393889-1; ISBN 13: 978-0-13-393889-0

17 2024

www.pearsonhighered.com

Praise for *Thinking in Physics*

"*Thinking in Physics* serves as a practical guide to teaching introductory physics and lays out evidence and motivation for the methods presented."
—*Steve Pollock, University of Colorado, Boulder*

"This is an extremely valuable resource for any physics instructor. *Thinking in Physics* not only provides the background research and tools necessary to assess student cognitive abilities and understand why some will likely struggle in introductory physics, but also introduces research-based pedagogical strategies and activities to boost student conceptual understanding and problem-solving abilities. It is a book every physics teacher should own and read."
—*Jeff Steinert, Arizona School for the Arts*

"In this book, Coletta makes the persuasive case that scientific reasoning ability is one of the main drivers of student success in learning physics. The book then describes a number of demonstrably effective strategies that you can use (in your class, tomorrow) to encourage this reasoning ability to develop. I applaud the work."
—*Nathan Moore, Winona State University*

"I was able to easily adapt the *Thinking in Physics* method to my physics course at Animo Venice Charter High School. In feedback surveys, students expressed greater engagement and confidence in learning science and my percentage of 'far below basic' and 'below basic' students decreased to the lowest ever."
—*Paul Hsu, Los Angeles Unified Schools*

"I am very impressed with the way Coletta utilizes evidence-based teaching innovations which develop both students' scientific reasoning skills and their learning in introductory physics. In this inspiring book, Coletta carefully describes all the pedagogical approaches and tools he uses, and shows the reader how and when they are incorporated in the course. Interestingly, while his approach clearly helps all students, it is exceptionally effective in closing the gender gap in conceptual gains between females and males. I firmly believe that physics instructors all over the world can benefit from the materials and methods presented in this book."
—*Antti Savinainen, University of Jyväskylä, Finland*

About the Author

Vince Coletta earned his Ph.D. in theoretical physics from the University of Notre Dame, where his graduate research advisor was Gerald Jones, an academic descendent of Uhlenbeck and Boltzmann. For over 30 years he has been a Professor of Physics at Loyola Marymount University in Los Angeles, where he teaches introductory physics and courses in electromagnetic fields, quantum mechanics, and statistical mechanics. In this book, Professor Coletta describes his *Thinking in Physics* curriculum and provides evidence for its effectiveness in developing students' scientific reasoning skills, understanding of physics concepts, and problem solving skills.

Preface

Do some of your students struggle unsuccessfully to understand the most basic concepts of physics? Does understanding physics seem impossible for them, despite your best efforts? It may be that their understanding is limited by a lack of strong scientific reasoning skills or by an inability to utilize those skills. This may be so even if you use interactive teaching methods that are far more effective than traditional instruction for most students. If so, this book is for you. It is the story of my struggle to reach students for whom interactive engagement* alone is not enough. After years of development, I have created a new curriculum, Thinking in Physics, that enables me to reach more of those students who have the greatest difficulty learning physics. This success has been achieved only after much trial and error, ongoing analysis of results, and refinement of methods.

In this book, I present evidence that students' reasoning ability is strongly related to their learning in physics and I describe ways for students to improve their reasoning and to achieve a better understanding of basic principles of physics. I provide sufficient details about my Thinking in Physics (TIP) course and enough sample materials so that if you want to teach your own TIP course, you can do so, adapting TIP to your own student population and to your own teaching style.

Although the focus of this book is physics, I believe that instructors in other STEM disciplines will find many of the ideas broadly applicable to their classes.

In Chapters 1 and 2, I describe the motivation for TIP, its development, and evidence for its effectiveness. In Chapter 3, I describe the important role that gender plays in learning physics. The research basis for TIP is provided in these first three chapters. Subsequent chapters describe the pedagogy in detail and provide materials. Mel Sabella describes in Chapter 7 the adaptation of some early TIP materials for courses at Chicago State University. Some additional materials are provided on the TIP website: ThinkingInPhysics.com.

My colleague Jeff Phillips has been my collaborator throughout the Thinking in Physics project, which was supported for several years by the National Science Foundation (063335). Together, we analyzed data and developed pedagogy. The details of the curriculum we each created are somewhat different, but very complementary. In this book, I present my version of TIP.

---

* Interactive engagement describes a broad range of methods, all of which have as their function the active engagement of students with concepts through interaction with other students or instructors.

I wish to express my gratitude to Jeff Phillips for working with me on the TIP project and for the contributions he made to this book, and to Joe Redish and Philip Adey, who were consultants on the project, for their valuable advice. Thanks to Mel Sabella for using early TIP materials in his classes at Chicago State, and for contributing Chapter 7 based on that experience. I am grateful for the help provided over the years by scores of teaching assistants and by my undergraduate research assistants: Michael Bruns, Giorgio Chirikian, Jonathan Evans, Raquel Sena, and Lisa Taylor. I am also grateful to the thousands of students who have responded positively to my efforts with them. I am indebted to the many teachers, authors, and physics education researchers whose work has helped and inspired me. These include Wendy Adams, Philip Adey, Arnold Arons, Tom Carter, John M. Clement, Catherine Crouch, Karim Diff, Michael Dubson, Dewey Dykstra, Eugenia Etkina, Reuven Feuerstein, Richard Feynman, Noah Finkelstein, Richard Hake, Charles Henderson, David Hestenes, Paul Hewitt, Brant Hinrichs, Paul Hsu, Tiffany Ito, Jane Jackson, Jerry Jones, Stephen Kanim, Anya Karelina, Robert Karplus, Lauren Kost-Smith, Nathaniel Lasry, Anton Lawson, Michael Loverude, Kandiah Manivannan, Eric Mazur, David Meltzer, Lillian McDermott, Nathan Moore, Katherine Perkins, Steve Pollock, David Pritchard, Brian Pyper, Joe Redish, Jeff Saul, Antti Savinainen, Chandralekha Singh, Jeff Steinert, and Carl Wieman. Thanks to Pearson publisher Jim Smith, who encouraged me to write this book, and thanks for facilitating the book's publication to the editorial and production staff: Becky Ruden, Katie Conley, Joe Vetere, Laura Kenney, and Sarah Kaubisch. Special thanks to David Hilsen, publisher of my textbook *Physics Fundamentals*, for granting permission to use figures and photos from that book, and to Matthew Peterson, co-founder of the MIND Research Institute, for allowing me to use the computer games he has created.

Vincent Coletta
Loyola Marymount University
Los Angeles, California

Contents

Preface   v

I   The Research Base for Thinking in Physics   1
     1   The Development of Thinking in Physics   2
     2   Evidence for the Effectiveness of Thinking in Physics   11
     3   Gender Effects   17

II   The TIP Curriculum   25
     4   Student Population, Institutional Constraints, and the TIP Classroom   26
     5   Introducing Students to Thinking in Physics   29
     6   TIP Course Structure   37
     7   TIP at Chicago State University by Mel Sabella   47

III   TIP Materials   51
     8   Connecting the Dots   52
     9   Reading Tests   56
     10   Clicker Questions   71
     11   Problem Solving Worksheets   118
     12   Labs and Lab Concept Quizzes   122
     13   Quizzes and Tests   185

Appendix 1: Lawson Classroom Test of Scientific Reasoning Ability   195
Appendix 2: A Guide to Learning Physics   203
Appendix 3: PowerPoint Slides for the First Week   211
Appendix 4: Syllabus and Schedule   217
Appendix 5: Seating Chart   221
References   222
Credits   225

Part I
The Research Base for Thinking in Physics

1 The Development of Thinking in Physics

2 Evidence for the Effectiveness of Thinking in Physics

3 Gender Effects

The Development of Thinking in Physics

"Science is a way of thinking much more than it is a body of knowledge."
Carl Sagan

The Thinking in Physics project evolved from the challenge of teaching introductory physics. In 1969, after completing a Ph.D. in theoretical physics at the University of Notre Dame, I began teaching at Loyola Marymount University in Los Angeles. Like many other new professors, I modeled my teaching on what worked for me as a student. My model was my thesis advisor, Jerry Jones, a great lecturer, a clear and rigorous thinker who cared deeply about his students' learning. I was also strongly influenced by two textbooks: *The Feynman Lectures*[1] and Hewitt's *Conceptual Physics*.[2] Though these two books are radically different in level, they have the common element of emphasizing qualitative reasoning. They convinced me of the value of focusing on student understanding of basic concepts, which I began to stress in my lectures and tests. I was committed to making my presentation of physics concepts more logical and accessible than the textbooks then available for teaching college physics, and so I began writing a textbook, initially using it in manuscript form with my students.

In 1990 Arnold Arons published his groundbreaking book *A Guide To Introductory Physics Teaching*.[3] In the Preface he states "My objective is to bring out as clearly and explicitly as possible the conceptual and reasoning difficulties many students encounter and to point up aspects of logical structure and development that may not be handled clearly or well in substantial segments of textbook literature." Arons' book describes in great detail conceptual problems related to relative motion, circular motion, free body diagrams, energy, and many other topics in introductory physics. I was deeply impressed by Arons' insights, which resonated with what I had observed in my own teaching experience. I incorporated many of Arons' ideas into my textbook. For example, wherever possible I introduced abstract concepts by appealing to common experiences that illustrate the concepts, and I devised questions and problems that emphasized conceptual understanding. My belief was that if I could just explain physics better, with a stronger emphasis on concepts, more students would get it.

In 1994 I attended a national meeting of the American Association of Physics Teachers (AAPT) at Notre Dame. There I became aware of the Force Concept Inventory[4] (FCI) and of the evidence for the effectiveness of nontraditional, interactive engagement methods of teaching physics. The FCI is a research-based multiple-choice test, with incorrect answers reflecting common student misconceptions about force and motion.*

After 25 years of teaching, I was excited by the prospect of having a test like the FCI to provide an objective measure of my teaching effectiveness. Use of the FCI and the choice of new alternative pedagogies provided a more scientific approach to teaching physics, making teaching physics more like doing physics. I was also pleased that the FCI stressed exactly the same basic concepts I emphasized in my teaching, and I was certain my students would do well on it. But when I gave the test to one of my classes just completing first semester physics, I was shocked by the poor results. Few of my students scored above 50%. I shouldn't have been surprised. My students' performance was consistent with results others had seen for courses using traditional teaching methods. Even though I emphasized qualitative conceptual understanding in my lectures and stressed conceptual questions in my new textbook[5,6] and on my tests, my classroom methods were entirely traditional. My students were not sufficiently engaged with the concepts to overcome their deeply entrenched preconceptions about force and motion and to master meaningful physics concepts.

I realized that I must change my approach. By 1996 I had developed an early version of my own interactive engagement (IE) class, in which topics are covered first in "concepts" classes, where students discuss conceptual questions with each other, and then again in problem solving classes, where problem solving strategies are developed. I designed a special classroom to facilitate the kinds of activities my students engage in, and I used the FCI to gauge the effectiveness of my new methods.

When the FCI is given both at the beginning and end of a course, average pre- and post-instruction scores can be used as a measure of conceptual learning achieved during the course. Normalized gain is a measure, introduced by Hake,[7] that allows one to compare classes with very different initial knowledge states. The class average normalized gain <g> is defined as:

$$<g> = \frac{\text{class average postscore\% - class average prescore\%}}{100\% - \text{class average prescore\%}}$$

Thus <g> is the actual change divided by the maximum possible gain. For example, the normalized gain of classes with average pre → post scores of 20% → 60%, 40% → 70%, and 80% → 90% all correspond to <g> = 0.5. Loosely speaking, <g> is the fraction of the concepts that a class learns that were not already known at the start of the course. Extensive research

---

*You can download the FCI from the Arizona State Modeling website – modeling.asu.edu. A password is required. Use your school's email to request a password from David Koch: FCIMBT@verizon.net

demonstrates that IE methods are consistently much more effective than traditional methods, as evidenced by much higher gains[7].

Soon after I began using IE methods to teach physics, my students' performance on the FCI improved, with normalized gains of about 0.4, consistent with results in many other IE classes. In 2002 I gave my first talk on my new interactive physics course. In preparation for this talk, I looked at the data Eric Mazur presented in his book *Peer Instruction*,[8] in which he describes the IE method he developed at Harvard. His method of emphasizing conceptual understanding through peer discussion was very similar to what I had independently developed in my concepts classes. I was therefore surprised to see that his students' FCI gains were about 0.6, much higher than the 0.4 gains of my own students, even though his class was focused on student discussion of concept questions that were similar in content and style to mine. Could it be that his gains were so much higher because of his elite student population, students who began his course with a much better understanding of physics concepts than typical college physics students? Certainly the Harvard average preinstruction FCI scores of about 70% far surpassed my students' averages of about 30%. Nationally most FCI class average prescores range between 25% and 50%.

Normalized gain was thought to be independent of the class average prescore. My thought was that normalized gains might actually be higher for classes with higher average preinstruction FCI scores. If this were true for entire classes, it should also be true for individuals. Perhaps students who start with higher preinstruction scores not only have higher postinstruction scores, but also have higher *individual* normalized gains, where the normalized FCI gain G for an individual student is defined as:

$$G = \frac{\text{individual postscore \% - individual prescore \%}}{100\% - \text{individual prescore \%}}$$

When I examined individual normalized FCI gains for my own students, I found a significant positive correlation between a student's FCI prescore and G ($r = 0.33$). Students with the lowest prescores (about 20%) had the lowest gains (averaging 0.3). Students with the highest prescores (about 70%) had the highest gains (averaging 0.6). I also began collecting student data from other universities – FCI prescores and postscores for individuals, generously shared by other professors: David Meltzer and Kandiah Manivannan provided data from Southeastern Louisiana University (SLU), Charles Henderson and the Minnesota PER Group provided University of Minnesota (UM) data, and Catherine Crouch and Eric Mazur provided Harvard data. The SLU and UM data were similar to my own data, with the lowest prescore students having Gs averaging 0.3 and the highest prescore students having Gs averaging 0.6. I also analyzed class average data for 38 IE classes and found a strong correlation between $<g>$ (class average normalized gain) and class average prescore ($r = 0.63$). Those classes with the lowest average prescores (about 30%) had <g>s that averaged 0.35 and those with highest average prescores (about 70 %) had <g>s that averaged 0.55. So all of these results supported my hypothesis that higher FCI prescores tend to produce higher normalized gains for both individuals and classes.

However, the Harvard data showed absolutely no correlation between prescore and normalized gain (Fig. 1.1). Even those Harvard students with very low FCI prescores, on average, had normalized gains of about 0.6, the same as classmates with much higher prescores. It occurred to me that it is something else about the Harvard population that leads to their success. Perhaps it was their ability to think scientifically, rather than their understanding of physics at the beginning of their first college physics course, that led to their unusually high normalized gains. So I began to look for a test that could measure such skills. My colleague Jeff Phillips suggested I look at a test measuring scientific reasoning skills that had been created by Anton Lawson, an Arizona State University biologist.

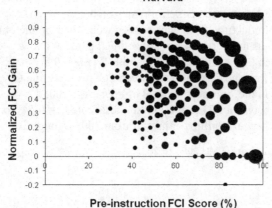

Fig. 1.1 At Harvard, individual students' normalized gain was not correlated with FCI prescore ($N = 670$) for Peer Instruction courses taught from 1991 to 1996. For example, the average of the individual normalized gains for the 20 Harvard students with the lowest prescores was 0.6, the same as the average G for students with higher prescores.

Scientific Reasoning Ability and Understanding Physics Concepts

According to Piaget,[9] an individual progresses through discrete stages of intellectual development, beginning with concrete and eventually reaching the formal reasoning stage needed to perform scientific reasoning – for example, the skill to control and isolate variables, and to search for relationships, such as proportions. Contrary to Piaget's notion that most teenagers have reached the formal stage, many high school and college students have not. Anton Lawson developed Lawson's Classroom Test of Scientific Reasoning Ability,[10] a multiple-choice test that includes questions on conservation, proportional thinking, identification and control of variables, probabilistic thinking, and hypothetico-deductive reasoning. A score of 75% or more on this test indicates fairly consistent formal reasoning. A score below 45% is an indication of concrete reasoning, and scores between 45% and 75% are transitional, neither fully concrete nor fully formal.

Fig. 1.2 shows an edited version of Lawson test questions on control of variables. The full test is provided in Appendix 1.

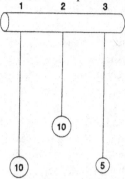

Fig. 1.2 Three strings have either 5 or 10 unit weights attached to their ends. Suppose you want to find out whether the length of a string has an effect on the time it takes to swing back and forth. Which strings would you use to find out?
a) only one string; **b)** all 3; **c)** 2 & 3; **d)** 1 & 3; **e)** 1 & 2. Because **a)** you must use the longest strings; **b)** you must compare strings with both light and heavy weights; **c)** only the lengths differ; **d)** to make all possible comparisons.

Using the Lawson test, Jeff Phillips and I found that among the engineering, biology and natural science students sampled at LMU, over 50% had not yet reached the formal level and consequently were ill-prepared for abstract reasoning in physics. We decided to investigate whether performance on the Lawson test, given at the beginning of an introductory mechanics course, would be a predictor of conceptual learning in physics, as measured by normalized FCI gain. What we found was a correlation between Lawson test prescore and FCI G ($r = 0.5$, $N = 65$) that was significantly stronger than the correlation between FCI prescore and G ($r = 0.3$). We published this result in the American Journal of Physics in 2005.[11] Fig. 1.1 shows the correlation between FCI G and Lawson test score ($r = 0.54$) for 98 LMU students taught by either Jeff Phillips or myself from 2003 through 2005.

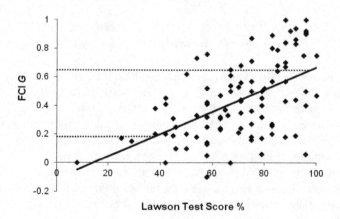

Fig. 1.3 Correlation between Lawson test preinstruction score and individual FCI normalized gain G for 98 LMU life science students from 2003 to 2005 ($r = 0.54$, $p < 10^{-4}$).

The relationship that we observed between scientific reasoning ability and conceptual understanding in physics has been widely replicated. Correlations between FCI normalized gain and Lawson test score very similar to those of my students have been observed by other researchers, including Jeff Steinert[12] at Edward Little High School in Maine, Steve Pollock[13] at the University of Colorado, Karim Diff[14] at Santa Fe Community College, Jeff Saul[15] at the University of Central Florida, Brian Pyper[16] at Brigham Young University, and Antti Savinainen[17] at Kuopion Lyseo High School in Finland. Though Eric Mazur did not do a correlation study, he did give the Lawson Test to one of his Harvard physics classes and he shared his results with me. The average score for the 165 students was 87%. Inspection of Fig. 1.3 shows that such a score predicts an average G of about 0.6, which is a typical value for $<g>$ in Mazur's classes.

In 2007 Jeff Phillips, Jeff Steinert, and I published a paper in *Physical Review*[18] showing that SAT scores (math + verbal) are correlated with normalized FCI gain, with a correlation coefficient of 0.5, close to the value for r between Lawson score and FCI G. Again the variation in G for groups of students with different SAT scores is quite large, ranging from about 0.2 for those whose SAT scores are about 800 to about 0.7 for those whose SAT scores are above 1400.

The strong correlations between SAT scores and FCI G and between Lawson scores and FCI G suggest that variation in reasoning abilities is a major cause of the observed wide range of class average normalized gains in IE classes across the country. Many courses use the same or equally effective pedagogies, but with students whose skills are quite different. Thus at Harvard, where most students have combined math and verbal SAT scores above 1400, normalized FCI gains in Mazur's Peer Instruction classes average about 0.6. But at Chicago State University, an inner city university whose students have ACT scores mostly below 21 (equivalent to SAT scores below 1000), $<g>$s in Mel Sabella's IE physics classes average about 0.3. For student populations like my own, with a wide range of SAT scores, mostly between 1000 and 1250, normalized FCI gains typically averaged about 0.4 before we developed TIP.

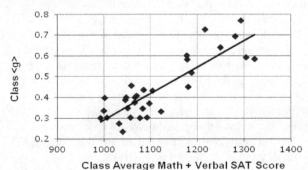

Fig. 1.4 Class average normalized gain $<g>$ is very strongly correlated ($r = 0.86$) with SAT score for the 31 classes taught by Jeff Steinert from 2000 to 2006.

Fig. 1.4 provides very compelling evidence of the effect of SAT scores on class average normalized gain when a single teacher uses the same methods to teach classes with widely varying average SAT scores. The figure shows data provided by Jeff Steinert, who used Modeling Instruction to teach 31 regular and honors classes at Edward Little High School from 2000 to 2006. (Modeling Instruction is one of the most effective and widely used IE methods, and Steinert is an expert in Modeling who often leads Modeling Instruction workshops.) The correlation coefficient between FCI <g> and SAT score is extraordinarily large: $r = 0.86$. Steinert's data show a similarly strong relationship between FCI <g> and class average Lawson test score.

The results of our research led to two obvious new research questions that we began investigating in 2005: 1) Can students' scientific reasoning ability be improved? 2) Can students with initially limited scientific reasoning ability achieve higher FCI gains than previously seen? Fortunately, the answer to both questions is yes. Students in a TIP physics course accomplish both.

Improving Reasoning Skills
We began the development of Thinking in Physics (TIP) by researching the efforts of others to improve reasoning skills. What we found were three programs dedicated to developing reasoning in very different populations. All of these programs were a source of inspiration and ideas for the development of TIP. The first source was Numerical Relationships (NR), created by Karplus and Kurtz[19] in the 1970s. NR activities were designed to help high school students with proportional reasoning, specifically distinguishing constant ratio, constant difference, and constant sum relationships. Instruction was provided during fifteen class hours over three weeks and was based on the Learning Cycle pioneered by Karplus.[20] Students who used the NR materials showed great improvement, while those enrolled in control sections did not. In developing TIP, we found that the same students who answered Lawson proportional reasoning questions incorrectly usually did quite well on a test that consisted of *only* proportional reasoning problems. Once these students recognized that the test was all about proportions, they knew how to solve those problems. Developing the ability to recognize and solve proportional reasoning problems is an important objective of TIP, and use of proportional reasoning is incorporated into many of the TIP activities.

The second TIP source was Israeli psychiatrist Reuven Feuerstein's Instrumental Enrichment (FIE).[21] Feuerstein, who was inspired by Piaget and Vygotsky,[22] developed the FIE program to improve the intellectual level of children with very low IQs. Feuerstein's success with these children confirmed his belief that intelligence is not fixed. The part of the FIE program that resonated with what we were trying to accomplish in TIP was its training to restrain impulsiveness. Impulsiveness is a common problem for the children Feuerstein worked with. It is also quite commonly a problem for students trying to do physics problems. Frustration and impatience often leads students to just grab an equation without a lot of thought about its applicability to the problem they are trying to solve. Impulsiveness can also lead to wrong answers on the Lawson test and on the FCI.

One of the most effective FIE activities involves connecting sets of dots to form simple geometric shapes. Students are first shown a picture of a few simple shapes, for example a square and two triangles. Then they are shown a second picture that appears at first to be a random array of dots (Fig. 1.5). When the dots are connected by the proper lines, they form the same shapes as in the first picture, a square and two triangles in our example. However, the shapes have been rotated and overlap, making the identification of shapes more difficult. There are many such sets of pictures of gradually increasing complexity. The exercise has the virtue of not relying on language, which is a barrier for many of the children Feuerstein worked with. Initially these children have great difficulty, because they tend to impulsively connect dots, which give rough approximations to the desired shape but are incorrect. Gradually they learn to restrain their impulsiveness, to pay attention

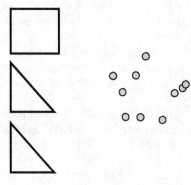

Fig. 1.5

to detail, and to develop strategies (such as find the square first), all of which are useful problem solving strategies in any context. TIP incorporates some of these dot exercises into the beginning of the physics course, and then continues the theme of restraining impulsiveness as an important part of the TIP approach to problem solving.

The third TIP source was Cognitive Acceleration through Science Education (CASE),[23] a program created by Philip Adey and Michael Shayer for enriching middle school science classes in Great Britain. Like Feuerstein, Adey and Shayer were inspired by the work of Piaget and Vygotsky. Their work is based on the belief that, though cognitive development is partly influenced by genetic make-up, it is also strongly influenced by cognitive stimulation, which instructors can produce in the classroom. The three pillars of CASE are cognitive conflict, social construction, and metacognition. Cognitive conflict is produced by observation of a surprising result. Under the right conditions this conflict can lead to an accommodation of the mind to a higher level, needed to deal with the greater demands made upon it. In order for this to happen, the social element must be present; that is, dialogue with others, teachers and students, is essential. Finally, students must be encouraged to reflect on their thinking and to transfer what they have learned to other domains.

CASE was implemented by inserting one hour lessons into the regular science curriculum once every two weeks for two years. These lessons all involved some kind of science content, but their purpose was to teach thinking. Specifically, the program taught identification and control of variables, proportional reasoning, probability and correlation, and use of abstract models.

One of the CASE activities is a study of floating and sinking. Students observe jars of various sizes and weights that either float or sink. Then they are shown a jar that is the same size as one of the other floating jars and is the same weight as another of the floating jars. They are asked to make a prediction about this new jar. After the prediction, they observe that the jar in question sinks. For most students this is a surprising result that creates cognitive conflict. The resolution of that conflict through discussion eventually leads to a formal level of understanding involving ratios and several variables. That is, one must compare the mass to volume ratio of the jar with the mass to volume ratio of the liquid in which it is immersed.

On average, students who participated in CASE, three years later averaged one full grade higher in science, mathematics, and even English than did students in control groups. CASE methods have been successfully replicated throughout the world.

From 2007 to 2010 I worked with my colleague Jeff Phillips on the development of TIP methods and materials, supported by a grant from NSF. Philip Adey served as an NSF consultant on the TIP project. He provided great advice and inspiration for this work. The CASE themes of cognitive conflict, social construction, and metacognition have been woven into the fabric of TIP. Since 2010 TIP has continued to evolve, addressing such issues as gender differences. In the next chapter I will present the evidence that TIP is working.

Essential to the TIP program is the overarching idea that intelligence is not fixed – that it is malleable. Although there is a genetic component to intelligence, programs like CASE, NR, and FIE, as well as modern research in cognitive psychology, demonstrate that intelligence is much less determinative than once thought. Cognitive research shows that achievement in science courses is strongly related to working memory and fluid intelligence,[24] and that working memory capacity is closely related to general fluid intelligence,[25] the ability to solve novel problems. Ravens Advanced Progressive Matrices[26] is a test often used to measure fluid intelligence. Jaeggi and others have shown that working memory can be improved through short term computer training[27,28] and the amount of improvement is greater for those who believe that intelligence is malleable, not fixed,[29] which is consistent with Dweck's finding that those who believe that intelligence is fixed tend to disengage from tasks perceived as too difficult.[30] Recently, changes in images of the brain have been seen to accompany training of working memory.[31] According to Buschkuehl, "These data suggest that the brain changes and becomes more physically fit as a function of training – a mental conditioning effect...." It now seems clear that both general fluid intelligence and specific scientific reasoning skills can be improved.

2
Evidence for the Effectiveness of Thinking in Physics

"If you want to teach people a new way of thinking, don't bother trying to teach them. Instead, give them a tool, the use of which will lead to new ways of learning."   Richard Buckminster Fuller

Conceptual Understandng of Physics
Between 1996 and 2007, after I had begun teaching physics using interactive engagement methods, but before I had implemented the TIP pedagogy, the results my students achieved on measures such as the FCI were very typical of interactive engagement classes comprised of students with similar reasoning abilities. For example, data supplied by Jeff Steinert for Edward Little High School (ELHS) in Maine showed distributions of Lawson scores, SAT scores, and normalized FCI gains for his students that were very similar to the distributions of those quantities for LMU students taught either by me or by my colleague Jeff Phillips. The three of us combined our individual student data to provide a clearer picture of how FCI G varies with Lawson scores and SAT scores.[18] Combined ELHS/LMU pre-TIP data are shown in Fig. 2.1. Because individual student data always show a wide variation in G for individuals with the same Lawson scores, data are grouped in bins. Each bin contains FCI G's for students with a small range of Lawson scores. The average value of FCI G for each bin is graphed vs. the bin's average Lawson score. Data from many other high school and college classes show a similar relationship between Lawson scores and FCI G and between SAT scores and FCI G.[13–17]

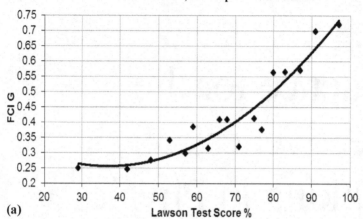

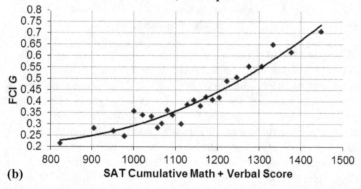

Fig. 2.1 (a) FCI normalized gain vs. preinstruction Lawson Test score for 297 LMU and Edward Little High School students, with individual student data averaged within 17 bins. (b) FCI normalized gain vs. SAT score for 627 LMU and Edward Little High School students, with individual student data averaged within 25 bins.

The graphs in Fig.2.1 provide a useful reference for measuring the effectiveness of a curriculum for developing understanding of the most basic concepts of force and motion in an introductory mechanics course. If you administer the FCI pre and post and either give the Lawson test at the beginning of the course or have access to SAT data for your students, you can compare results in your own course with one of these graphs. So, for example, if your class average Lawson test score is about 70% or if your class average SAT math plus verbal score is about 1150, then these graphs suggest that a FCI normalized gain of about 0.4 might be expected if IE methods are used. If your students' results are significantly above the trendline on either graph, then your students are doing better than this reference level. Fig. 2.1 has provided me with an ongoing means of gauging the success of my new TIP pedagogy.

With the most recent version of the TIP curriculum, more of my students now attain a stronger understanding of physics concepts than in my earlier IE courses[32] that preceded TIP and provided the data for Fig. 2.1. Fig. 2.2 shows the results for the latest version of TIP (2012 to 2013), with the trendline from Fig. 2.1a, representing the earlier pre-TIP data, included for easy reference. TIP students with the lowest Lawson scores showed the biggest improvement in FCI gains, compared with pre-TIP classes. Average Gs for all of the bins except the highest two are two or three standard errors above the reference trendline. Thus Fig. 2.2 shows that many TIP students whose initial scientific reasoning skills are such that, before TIP, they would have had very low FCI gains are now able to achieve much higher gains, indicating a substantial conceptual understanding of physics.

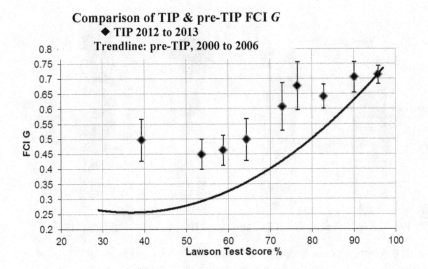

Fig. 2.2 FCI normalized gains binned by Lawson test scores for TIP students, 2012 to 2013 ($N = 76$). The trendline represents pre-TIP LMU/ELHS data from 2000 to 2006. Error bars indicate standard errors. The lowest 3 bins are about three standard errors above the trendline and the next highest four bins are about two standard errors above the trendline.

I made significant changes to the TIP curriculum from 2007 to 2012. As I did, my students' postinstruction FCI scores and normalized gains improved (Fig. 2.3). Because the characteristics of a student population can have a dramatic impact on the success of any curriculum, it is important to consider whether my student population changed significantly over the years of this study. My student population has been very stable from 2000 to 2013. All classes were introductory mechanics for life sciences, and for each of the groupings shown in Fig. 2.3, the percentage of females was between 60% and 66%, the average preinstruction Lawson score was between 67% and 71%, and the average preinstruction FCI score was between 27% and 29%. Much of the improvement in FCI scores from 2010 to 2012 has been due to dramatic improvement in the FCI gains of my female students over this period, as I will explain in the next chapter.

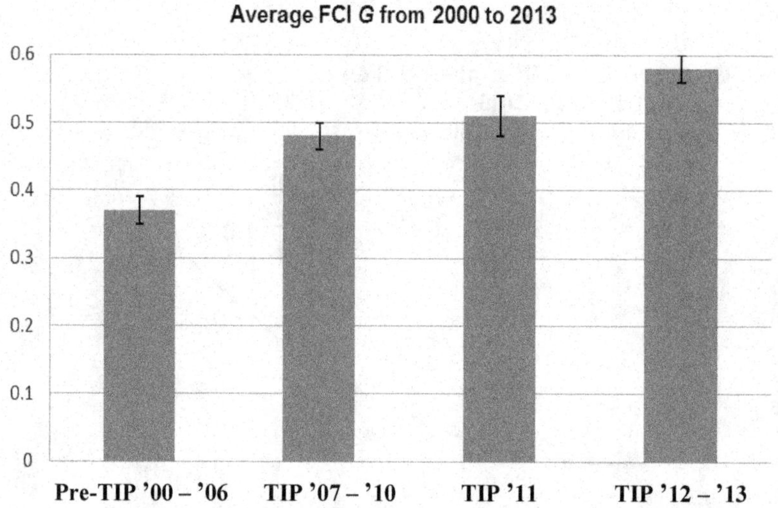

Average FCI G from 2000 to 2013

[Bar chart showing average FCI G values: Pre-TIP '00 – '06 ≈ 0.37; TIP '07 – '10 ≈ 0.48; TIP '11 ≈ 0.51; TIP '12 – '13 ≈ 0.58. Error bars indicate standard errors.]

Fig. 2.3 Average normalized FCI gains steadily improved from 2000 – 2006 to 2012 – 2013. Numbers of students for each interval varied from 58 to 104: '00 – '06, $N = 101$; '07 – '10, $N = 104$; '11, $N = 58$; '12 – '13, $N = 76$. The percentage of females varied from 60% to 66%: '00 – '06, 66%; '07 – '10, 60%, '11, 66%; '12 – '13, 63%. Average Lawson prescores varied from 67% to 71%: '00 – '06, 68%;* '07 – '10, 71%; '11, 67%; '12 – '13, 70%. Average FCI prescores varied from 27% to 29%: '00 – '06, 29%; '07 – '10, 29%; '11, 28%; '12 – '13, 27%. Error bars indicate standard errors.

---

*From 2000 to 2006, only 41 of the 101 students who took the FCI also took Lawson's Test. This subset had an average Lawson prescore of 68%. For all other periods, all students took both the FCI and Lawson's test.

Improving Scientific Reasoning Skills

A basic objective of the Thinking in Physics curriculum is to develop students' scientific reasoning abilities along with their understanding of physics concepts. The Lawson test is given at the beginning and end of the first semester of physics. From 2012 to 2013 the average pre and post instruction Lawson scores were 70% and 80% respectively. The change in Lawson scores by prescore quartiles (Fig. 2.4) shows that the improvement in Lawson scores was greatest for those whose prescores were lowest. Gains were significant for the lowest three quartiles (47% → 68%, 63% → 73%, and 77% → 87%).

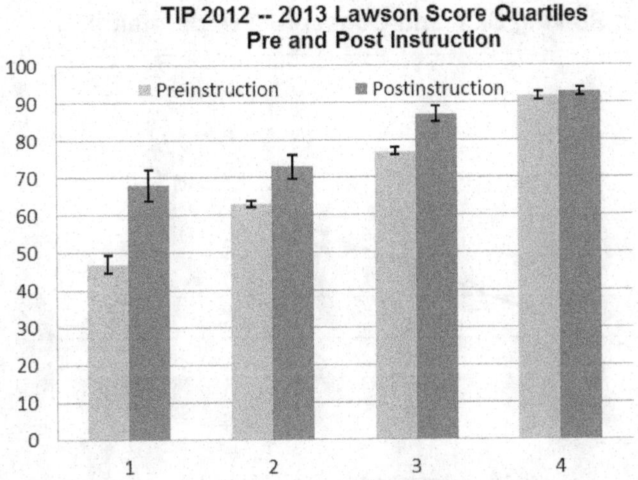

TIP 2012 -- 2013 Lawson Score Quartiles Pre and Post Instruction

Fig. 2.4 Pre and post instruction Lawson scores, divided into Lawson prescore quartiles, TIP 2012 – 2013 ($N = 76$). Error bars indicate standard errors.

Improving Problem Solving Skills

Developing strong problem solving skills has always been an important part of my teaching agenda. Some may question whether emphasis on conceptual understanding comes at the expense of diminished ability to solve problems. Nothing could be further from the truth. A strong conceptual foundation strengthens problem solving. Even though less time may be spent solving example problems in class, students' improved understanding of concepts enables them to be better problem solvers. A deep understanding of concepts makes it less likely that students are simply following an algorithm without a real understanding of what they are doing. When students encounter a really challenging problem, one that is not very similar to problems they have seen before, they are more likely to be successful if they have a strong conceptual foundation. PER data support this argument. For example, Eric Mazur showed that his students' problem solving improved after he introduced his Peer Instruction pedagogy, with its heavy conceptual emphasis. He gave the same final exam before and after making his change in teaching emphasis and found that the average score improved from 63% to 69%.

The Mechanics Baseline Test (MBT)[33] is a standardized test sometimes used as a measure of problem solving ability. Evidence that improved conceptual understanding leads to improved problem solving was provided in 1998 by Richard Hake,[7] who collected class average MBT and FCI data from high school and college IE introductory physics classes. Fig. 2.5 shows the strong correlation between students' post instruction performance on the MBT and the FCI in these classes ($r = 0.86$).

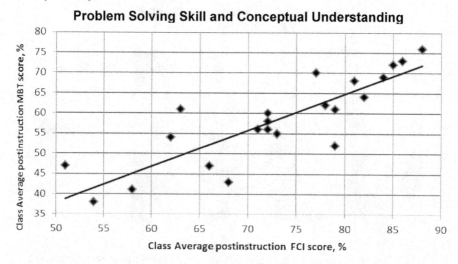

Fig. 2.5 Problem solving skill is strongly correlated with conceptual understanding in introductory mechanics, as evidenced by a very strong correlation ($r = 0.86$) between class average postinstruction scores on the Mechanics Baseline Test, used as a measure of problem solving skill, and class average postinstruction scores on the Force Concept Inventory, which measures conceptual understanding. Data from 24 high school and college IE classes (Hake,[7] 1998).

I have given the MBT at the end of each of my introductory mechanics classes since 1998 in order to gauge my students' problem solving ability. MBT scores improved after I began the TIP project. The average MBT score from 1998 to 2006 was 46% ($N = 171$), while the average score from 2007 to 2013 was 54% ($N = 251$). Standard errors for both were 1%.

3
Gender Effects

"We must have perseverance and above all confidence in ourselves. We must believe that we are gifted for something."   Marie Curie

Gender, Academic Testing, and Stereotype Threat

Each year over one million college bound high school seniors take the SAT, and each year the average score for men on the math section of this test is over 30 points higher than the average for women[34]. The math SAT gender gap has decreased somewhat in recent years: in the 1980's the difference between male and female scores averaged 42 points, in the nineties the difference averaged 37 points, and from 2000 to 2012 the difference averaged 32 points. Though there has been a slow decrease in the size of the gender gap, there remains a substantial difference between the average math SAT performance of men and women. The difference is greatest among the highest performing students. In 2012, 10% of all men, but only 5% of all women, scored 700 or higher. Researchers have puzzled for many years over this math SAT gender gap.

In 1995 Steele and Aronson[35] introduced the concept of stereotype threat: a person's feeling of being at risk of confirming a negative stereotype about a group to which the person belongs when that person is placed in a stressful situation that has the potential to confirm the validity of the stereotype. Steele and others have presented evidence that stereotype threat can account for much of the relative underperformance of females on the math SAT. A proposed explanation for this phenomenon is that a female test taker's concern that others expect her to underperform relative to males in math (the stereotype) brings up emotions that require regulation, thereby depleting the limited resources of her working memory, which leads to weaker performance on the test.

The FCI Gender Gap

A gender gap in student performance on the FCI has been widely observed, with males achieving consistently higher average normalized gains than females. Although some have speculated that the test itself is gender biased, Osborn Popp, Meltzer, and Megowan-Romanowicz[36] have presented evidence that the test is not systematically biased in favor of males. This suggests that the observed gap represents a real difference between males' and females' conceptual understanding of introductory mechanics.

For many years I was unaware that a gender gap existed in my own students' FCI scores. In terms of grades, women have always done about as well as men in my classes, and until very recently I never looked for possible gender differences in my FCI data. In 2011 I finally did look at my data with respect to gender, and I was very surprised to learn that not only was there a gender gap, but that the gap was quite large and that the gap was greatest among those men and women with the highest SAT and Lawson scores.[37] Once I began to look for causes and devise solutions, my female students' scores began to improve.

Others have reported an FCI gender gap based on overall class average results, typically with a difference in normalized gain on the order of 0.1. For those classes I taught from 2000 to 2006, after I had developed my basic interactive methods, but before I had developed the TIP curriculum, the difference in men's and women's average FCI G was over 0.2 (0.29 for females vs. 0.52 for males). Although the gap was very large and had persisted for quite a long time, somehow I was unaware of it. I think it was a matter of not wanting to believe that there was any difference between men and women in my classes, and therefore not looking. Being blind to what one sees due to preconceptions is not uncommon; we see it all the time in students' observation of physical phenomena in the laboratory.

This gender gap was far from uniform across different levels of reasoning skill. The gap was much larger for students with stronger reasoning skills than for those with weaker skills, as seen in Fig. 3.1. At the upper ends of the trendlines, the difference in normalized gains is 0.3, representing a huge difference between male and female FCI postinstruction scores: an average FCI score of 52% for high SAT females vs. an average of 86% for high SAT males. The data indicate that even before TIP, for many of the highest SAT male students, my interactive teaching methods were working quite well, but these same methods were a dismal failure for many female students, including those with high SAT scores. There were of course always individual exceptions. One of my female students with an SAT of 1400, scored 93% on the FCI, postinstruction. Yet most females with high SAT scores had relatively low postinstruction FCI scores. Data from ELHS show a very similar pattern (Fig. 3.2). Both LMU and ELHS data show a large gender gap for students with SAT scores above 1200. A similar gap in FCI G is seen for Lawson scores above 75%. This phenomenon is also seen in data from other schools, though the size of the gap at the upper end varies. This effect is consistent with the gender difference among high performing students on the math SAT.

When students answer an FCI question incorrectly, often they are not completely unaware of the physics principle that applies. It is just that their preconception is too powerful for them to resist when it is appealed to in this way. I have occasionally asked an individual student to take the FCI a second time in my office, to read aloud the question, and to express out loud thoughts about it before choosing an answer. Sometimes this different way of taking the test can produce a very different outcome. One student who had initially scored only 10 out of 30 on the FCI scored 20 when she talked her way through it.

Gender Effects

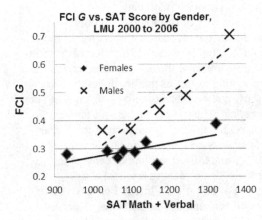

Fig. 3.1 Among students in my IE physics classes between 2000 and 2006, a large gender gap was observed for those students with SAT scores over about 1200. Males with the highest SAT scores had *Gs* about 0.3 higher than females with similar scores. Data are binned (8 bins for 63 females and 5 bins for 28 males); the standard error for each bin varies from 0.03 to 0.10.

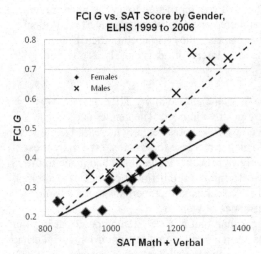

Fig. 3.2 Among students in Jeff Steinert's Modeling physics classes between 1999 and 2006, a large gender gap was observed for those students with SAT scores over about 1200. Males with the highest SAT scores had *Gs* about 0.25 higher than females with similar scores. Data are binned (13 bins each for 161 females and 171males); the standard error for each bin varies from 0.03 to 0.08.

I examined the subset of the ELHS data corresponding to Lawson scores of 42% or less (10 or less out of 24), with the scores of 14 females and 7 males falling in this range. For those students with the lowest Lawson scores, I observed a **reverse gender gap** on the FCI, with females achieving higher gains than males. The females with scores of 42% or less had an average FCI *G* of 0.30 ± 0.05 (SE) and the males with scores in this range had an average FCI *G* of 0.14 ± 0.04 (SE). I believe that it is likely that in most classes with very low Lawson scores (averaging 40% or less) or very low SAT combined math and verbal scores (averaging 800 or less) one will observe a reverse gender gap for the entire class.

Addressing Stereotype Threat at the University of Colorado

In November 2010, researchers at the University of Colorado (CU) published a paper in *Science*[38] in which they argued that stereotype threat is a cause of the gender gap they observed in test results in CU physics classes. They described an intervention they had designed to eliminate stereotype threat. The intervention consisted of two brief self affirmation writing exercises that I will describe in Chapter 5. The theory is that the self affirmation protects students from being preoccupied with negative stereotype thoughts. The CU researchers presented evidence showing that the intervention resulted in the elimination of the gender gap. Among those participating in the intervention, female and male students performed equally well on the FMCE (a standardized mechanics concept test used at CU instead of the FCI). When I first saw this research presented at a conference,[39] my initial reaction was one of disbelief. Though I believed that stereotype threat was a very real phenomenon, one that could perhaps account for much of the FCI gender gap, I found it very difficult to believe that the writing exercises could have the dramatic effect of eliminating the threat and the gap. And yet that is exactly what happened at CU. I was motivated to try the intervention myself. The authors of the CU paper were very helpful, especially Steve Pollock and Tiffany Ito, who provided all the materials they had used in the CU intervention, so that I could replicate their intervention as closely as possible.

Refining the TIP Curriculum

In 2011, I began using the Colorado writing exercise in my TIP classes. I was happy to find that it seemed to have a positive impact. The gender gap was reduced, though not eliminated. Average FCI scores actually increased somewhat for both men and women, but with a bigger increase for women, especially those with high Lawson prescores. The only significant change that I had made to the course in 2011 was the use of the two 15 minute self affirmation writing exercises. I felt that if this small change could have such an effect, perhaps there was more that I could do to improve the performance of women.

My original gender gap was bigger than that at CU. Stereotype threat is a kind of anxiety, and anxiety is often present in a challenging test environment. Perhaps my course created even more anxiety among my students than occurred in the CU course, with women being more susceptible than men. When I questioned my students about their level of anxiety about taking physics, I found that it was pervasive. Three fourths of my students had significant anxiety at the beginning of the course, with significantly more women than men reporting anxiety. Research[40] has shown that anxiety over academic performance is more common among women than men.

I was also concerned about students not fully understanding all that I was trying to accomplish in my course. Some comments from student evaluations of teaching had indicated this. For years I had struggled with the question of how much I should share with my students about the goals and methods of my course. Should I talk about reasoning skills and how theirs would be improved in this course? Some students might not respond well to this.

Student course evaluations also revealed that most of my students had no real interest in physics. They were taking the course only because it was required. Interest in physics is generally significantly stronger among males than females,[41] so addressing the gender gap in interest might help to close the FCI gender gap.

I decided that in order to address these problems, I needed to do more than simply give the writing exercise, which of course I continued to do. I needed to do whatever I could to reduce the level of anxiety, to increase understanding of my teaching methods, and to increase my students' interest in physics. To meet these goals, beginning in 2012, I introduced the following changes to my course:
1) a group project; **2)** online reading tests; **3)** a learning guide; **4)** interviews.

1) The purpose of the group project is to get students involved in something they are already interested in, to get them to apply physics to something – anything – of interest to them. Each group creates a 5 minute YouTube video in which they show the class how the physics they have learned applies to something they (and likely others) are already quite interested in.

2) The online reading tests provide students with an opportunity to take their first test on each new subject in an environment in which anxiety is minimized. Each online reading test consists of five multiple-choice questions based on an assigned reading from the text, and is taken by students anytime before the lecture on the material covered in the reading. Using the Blackboard system, students are given three chances to answer the questions.

3) I prepared a five page "Guide to Learning Physics," in which I explained in great detail course objectives, methods, the rationale for those methods, and evidence for their effectiveness. I also offered detailed advice about how to study, utilizing the various elements provided in the course. From the very beginning of the course, students now have a better understanding of the rationale for course methods and are better prepared for the challenges the course will present.

4) I now interview each of my students for 15 minutes during the first week of class. The interviews follow testing with the FCI, Lawson's test, an attitude survey, and submission of other personal information. I explain that the purpose of the interview is to get to know each of them and to make them more comfortable coming to see me for help with physics when they might need it. Establishing a personal relationship with my students during the first week allows me to understand who they are and makes them feel more comfortable about the course and about talking to me.

My hope was that these new course elements (described in more detail in chapters 5 and 6) would improve learning, especially among women, and indeed this was the case. In the data that follow you will see that in 2012 and 2013 TIP students, especially women, were more successful than ever before.

Reducing the FCI Gender Gap in TIP Classes

Fig. 3.3 shows the improvement in my students' FCI gains and the gradual reduction in the gender gap from pre-TIP classes (2000 to 2006) to the latest version of TIP (2012 and 2013). From 2007 to 2010 my classes had improved FCI gains for both females and males and some narrowing of the gender gap. In 2011 the gap was further reduced. Refinement of the curriculum in 2012 and 2013 resulted in substantial gains for both men and women and further closing of the gender gap: average G of 0.54 for females vs. 0.64 for males. Most significantly, the gender gap among students with the top half of Lawson scores, where previously the gap had been greatest, now has completely disappeared.

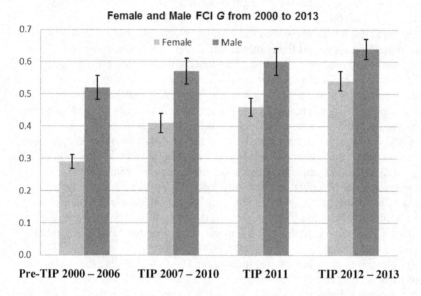

Fig. 3.3 As instructional methods have evolved, normalized FCI gains have improved for both females and males. The improvement has been greatest for females, with average female FCI G now greater than the average pre-TIP male gain. A gender gap still exists, but is now roughly half as great as it was in pre-TIP classes. The number of female and male students (N_F, N_M): preTIP (67, 34), TIP 2007 – 2010 (62, 42), TIP 2011 (38, 20), TIP 2012 – 2013 (48, 28). Error bars indicate standard errors.

Fig. 3.4 shows how FCI gains improved from 2007 to 2012/2013 for male and female students with different Lawson prescores. From 2007 to 2010 there was still a very large gap between high Lawson score males and females, as there had been before TIP. By 2012/2013, there was no significant gender gap among high Lawson score students. The biggest improvement was for high Lawson score females and low Lawson score males.

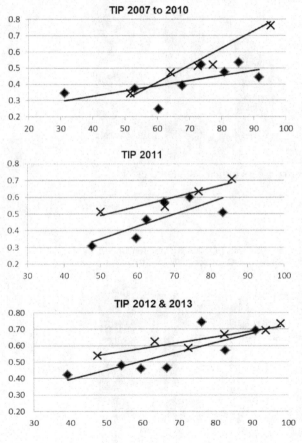

FCI G vs. Lawson Preinstruction Score (%) by Gender
♦ Females   × Males

Fig. 3.4 For 2007 – 2010, the number of females N_F = 63, the number of males N_M = 42; standard errors for female bins ranged from 0.05 to 0.08 and for male bins ranged from 0.04 to 0.10.
In 2011, N_F = 38 and N_M = 20; standard errors were between 0.05 and 0.06 for female bins and between 0.05 and 0.10 for male bins.
For 2012 – 2013, N_F = 48 and N_M = 28; standard errors were between 0.05 and 0.09 for female bins and between 0.02 and 0.10 for male bins.

Improvement in Lawson Scores by Gender

As TIP has changed, postinstruction scores on the Lawson test improved for both males and females. Using separate male and female class average normalized gains, we find that from 2007 to 2010, LT $<g>$ = 0.28 for females and 0.33 for males, in 2011, LT $<g>$ = 0.31 for females and 0.46 for males, and from 2012 to 2013, LT $<g>$ = 0.36 for females and 0.28 for males. Figs. 3.5 and 3.6 show female and male pre and post Lawson scores for prescore quartiles.

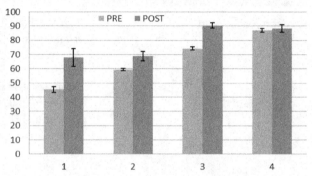

Fig. 3.5 Error bars indicate standard errors.

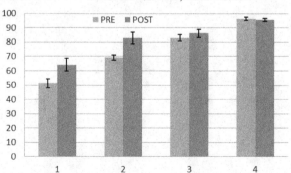

Fig. 3.6 Error bars indicate standard errors.

Part II
The TIP Curriculum

4 Student Population, Institutional Constraints, and the TIP Classroom

5 Introducing Students to Thinking in Physics

6 TIP Course Structure

7 TIP at Chicago State University

4

Student Population, Institutional Constraints, and the TIP Classroom

"If we teach today's students as we taught yesterday's, we rob them of tomorrow." John Dewey

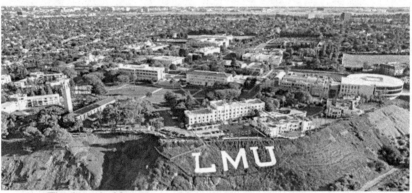

The Loyola Marymount University campus in Los Angeles

The Thinking in Physics curriculum was designed to meet the challenge of teaching introductory physics to life science students at Loyola Marymount University (LMU), a Jesuit university in Los Angeles, with an undergraduate enrollment of 6,000. As a private university, LMU is fortunate to have the resources to be able to limit class size, providing the opportunity for faculty to interact with individual students more often than would be possible if classes were much larger. LMU is a selective private university, but not a highly selective elite university. The student population does not include those with SAT scores as low as some who take introductory college physics courses. Yet less than half the students have ideal reasoning skills for the study of physics. Combined math and verbal SAT scores typically range from about 1000 to 1400, with most students in the range from 1000 to 1200. The diversity of student skills, comparable to the skills of those who take physics at a suburban public high school, presents a challenge. Significant time must be devoted to developing the scientific reasoning skills of some students. I typically teach either one or two sections of introductory physics for life science students, with the total number of students ranging from 20 to 60.

After several years of using interactive methods to teach introductory physics in traditional classrooms, I designed a new classroom that would facilitate interactive methods and convinced the University to build it. By 2003 the LMU Interactive Physics Classroom (IPC) was completed at a cost of about $150,000, including equipment. LMU introductory physics classes for both life science majors and engineers are taught in the IPC in sections of 32 students or less. The IPC has eight tables, each with seating for four students (Fig. 4.1).

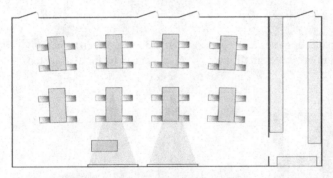

Fig. 4.1 Layout of the Interactive Physics Classroom.

Students are seated facing the front of the room, where instructors' presentations are facilitated by a dual screen computer connected to two projectors. A control panel allows for each projector to display either computer screen, or to display the input to a high resolution document camera, allowing documents or small demonstration objects to be viewed (Figs. 4.2 and 4.3). Projector screens are mounted above blackboards covering the front wall of the classroom. Glass panels extending out from the sides of each table provide writing surfaces. Mounted beneath each glass panel is a 17 inch flat panel computer monitor. The computer keyboard is mounted on a slide out tray, and the computer is stored beneath the tabletop (Fig. 4.4). Thus each student has a clear view of presentations at the front of the classroom, can use a computer that does not obstruct sight lines, and, by swiveling a chair 90°, can engage in a discussion with a partner across the table. The presence of a second pair of students at the table makes it easy to extend the discussion to a group of four. Students are provided with clickers attached to each work station (Fig. 4.5). When seating is assigned, as it is in my classes, each clicker can be associated with the student assigned to the seat, and one can keep track of responses. The class response to a clicker question is displayed as a bar graph showing the distribution of answers for the class to see; neither the students nor the instructor sees any individual's response. A grid shows when each student responds, thereby ensuring 100% participation.

Fig. 4.2 Professor Phillips uses a document camera to project a small demonstration.

Fig. 4.3 Dual screen monitor displays can both be projected.

Fig. 4.4 Student work station with computer monitor viewed through a glass writing surface.

Fig. 4.5 Partners respond to a clicker question.

The TIP curriculum described in this book is tailored to small class sizes and students with the range of reasoning skills of LMU students, and so significant adaptation is required if one wishes to use the TIP curriculum with much larger classes or with students who have different levels of reasoning skills. Chapter 7 describes an adaptation of TIP to students at Chicago State University.

5
Introducing Students to Thinking in Physics

"The mind is not a vessel to be filled, but a fire to be kindled." Plutarch

The first week of TIP is critical to the success of the course. I devote this week to explaining learning objectives and methods, discussing a written guide to learning physics, and getting to know the students as individuals through assessment and interviews (Table 5.1). Powerpoints for the first week of classes are provided in Appendix 3.

First class 75 minutes	Force Concept Inventory (25 min) Intro to Thinking in Physics (50 min)	Homework: read syllabus and guide to learning physics, submit first thinking journal entry, and schedule interview
Second class 75 minutes	Colorado self affirmation exercise (15 min), Introduction to problem solving (35 min), Lawson test (25 min)	Homework: first reading assignment from text and first online reading test, to be completed before the next class
First lab 110 minutes	Attitudes Survey (10 min) Skill building games	
15 minute interviews (20–60 students) 5 – 15 hours	Discuss assessment results; address anxiety, concerns	

Table 5.1 Schedule for the first week of class

Music Before Class
Throughout the semester, five to ten minutes before each class, I play a YouTube classical music selection. Some students enjoy this, though I suspect most would prefer that I play something else. I do it because I believe it sets a certain desirable tone, and also because I love the music I play and this puts me in an incredibly good mood. I play Moonlight Sonata before the first class – soothing sound to start the year.

The First Class
I begin the first class by giving the FCI, explaining that I use this test to find out what students already know about physics. The remainder of the class is devoted to an introduction to the Thinking in Physics course. I begin by asking the students a clicker question about their interest in physics (Table 5.2, question 1). Typically only about one third of the students say they have a strong interest in physics. They are life science majors and the course is required, so this is not too surprising. Next I ask whether they have anxiety about taking the course (Table 5.2, question 2). About two thirds of the students say yes. Seeing the class responses to these two questions can help individual students, as they realize that their anxiety and lack of strong interest are shared by most of their classmates.

#	Clicker Questions	Answers
1	Do you have a strong interest in learning physics?	A) Yes(30%); B) No (70%)
2	Do you have anxiety about taking this course?	A) Yes (70%); B) No (30%)
3	Is a force required to keep an object moving?	A) Yes (close to 100%); B) No
4	Will your concepts change as a result of having correct concepts explained to you?	A) Always (20%); B) Usually (40%); C) Usually not (40%)
5	Can you learn physics by only paying close attention while everything is carefully explained to you?	A) Yes; B) No (close to 100%)
6	If your overall average at the end of this course were 80%, what would be your grade in this class?	A; B; C Correct answer is A; most answer B.

Table 5.2 Clicker questions for the first class and a typical distribution of answers.

The first clicker questions provide a lead-in for me to discuss motivation and anxiety, and how students can deal with these issues. I encourage students to try to develop an interest in physics because this will make the hard work required in the course easier. I point out that physics can transform the way they see the world. I compare someone who views the world without a knowledge of physics to one who is completely color blind, a monochromat who sees the world in black and white and gray. When a person learns physics, it opens up a new view of the world, as though the monochromat begins to see the world in color for the first time. I promise students that what they learn is going to relate to all kinds of phenomena in their everyday lives. The point I try to make is that they should be open to the experience.

I suggest that what makes physics challenging for so many students is that it requires change: changing concepts about the physical world, changing the way they approach problem solving, and even changing the way they think about learning. I illustrate the change in physics concepts by asking whether a force is required to keep an object moving (Table 5.2, question 3). After nearly everyone indicates that they believe a force is required to keep an object moving, I pass out a toy hovercraft to each table and ask students to turn on the hovercraft so that it floats on a cushion of air and give the hovercraft a push, and then to discuss the question just posed in light of what they observe. Remarkably, when the same question is asked again after the experiment and discussion, a significant number of students still respond with a yes. After a little further discussion students all come to the conclusion that the correct answer is no.* I then point out that everyone has some concepts related to force and motion that work well for everyday life, but which have limitations, and that these concepts are remarkably resistant to change, as they have just witnessed.

Next I turn to a discussion of how to learn physics. I begin by asking whether they can learn physics concepts just by having them explained (Table 5.2, question 4). By now some students are catching on: few answer "always;" roughly equal numbers answer "usually" and "usually not." After some discussion of this question, they vote again, with some increase in the usually not's. I then explain that at one time I would have answered either "always" or "usually," and I based my teaching on that misconception; that is, I gave traditional lectures. I point out that a lot of research now shows that the correct answer is "usually not," and that is why I have changed the way I teach, with a lot more student engagement and discussion. I finish this theme by asking whether students can learn physics by only paying close attention to explanations in lecture (Table 5.2, question 5). A large majority answer "no."

Life science students are strongly motivated by grades. I begin a discussion of the course grading system by asking students what grade they think would correspond to an 80% average in this course (Table 5.2, question 6). Most guess that 80% would mean a B, and they are surprised to hear that the correct answer for this course is A, and that the minimum averages for B's, C's, and D's are roughly 70%, 60%, and 50% respectively. What this means of course is that tests and quizzes are going to be challenging. I believe that testing in this course should be roughly as challenging as the diagnostic tests like the FCI, the FMCE, or the Mechanics Baseline Test.

---

*I give the FCI at the beginning of class rather than the end so that students' FCI answers are not influenced by this discussion.

Next I give students a syllabus that provides a breakdown of points for the various course elements – tests, labs, etc. and a complete schedule of classes, labs, and homework assignments (Appendix 4). I also hand out a written Guide to Learning Physics (Appendix 2) and spend the remainder of the first class going over it with them. Students are later asked to read and comment on it.

A Guide to Learning Physics

The guide explains in great detail course objectives, methods, the rationale for those methods, and evidence for their effectiveness. I explain how students' reasoning skills, their conceptual understanding of physics, and their problem solving ability will improve. I specifically describe the role that various elements of the course play in accomplishing those goals: textbook reading, classroom clicker questions, homework problems, special skill building games, metacognitive thinking journals, labs, tests, and grading. I also offer detailed advice about how to study, utilizing the various resources provided in the course. I explain that much of what they learn in physics will be counterintuitive, and that the way they learn in this class may also be counterintuitive, very different from the way they may be used to learning. I explain that if I were to give them a traditional lecture course in physics, such a course would not be effective, and explain why it wouldn't work. I warn that the course will be very demanding, but that they, like many others before them, can succeed, and I promise to do everything I can to help them. I present data that show how well students typically learn in this course – evidence of success that justifies my methods. I use vague language to describe outcomes: "understanding of physics concepts, %" and "scientific reasoning skills, %", rather than saying that these percentages are scores on the FCI and Lawson's test (Fig. 5.1). I inform them that in a recent class over 50% of the students ended the course understanding >70% of physics concepts, but that one out of seven understood less than 50% of the concepts. So students are shown that success is common, but not guaranteed. Success requires a strong effort.

I complete the first class by showing some anonymous student comments about past courses, for example: "hardest class I've ever taken," "forced me to think from every angle," and "very difficult but learned so so much. Loved learning about physics." Some comments offer advice to new students, e.g. "Reading the text carefully, multiple times is very important. Also, when doing problems, don't just go through the motions, but think about why you do each step." "Understanding concepts is actually more important than doing problems because it shows the whole picture and will eventually help with problem solving."

From the very beginning of the course, students have a good understanding of the rationale for course methods and are prepared for the challenges the course will present.

Introducing Students to Thinking in Physics

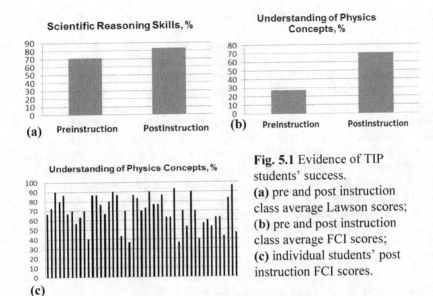

Fig. 5.1 Evidence of TIP students' success. (a) pre and post instruction class average Lawson scores; (b) pre and post instruction class average FCI scores; (c) individual students' post instruction FCI scores.

Thinking Journals

Metacognition is important for success in learning physics. Awareness of one's thinking is the first step to improving thinking. Students receive a small amount of credit for submitting thinking journal entries, based on prompts I provide periodically throughout the semester. For organizational convenience, their entries are Word documents placed in a "dropbox" I share with them, provided by the Dropbox website. The following are the first set of prompts given at the end of the first class meeting: (1) Tell me something about yourself. (2) Have you had a good high school physics course? (3) What are your goals for yourself in this course? (4) How much time each week do you plan to study outside of class? (5) Describe your experience with a concept or belief that you once thought was correct and no longer do, and how that came to change. (6) What are your thoughts on the course objectives and learning guide?

The Second Class

The second class begins with the Colorado self-affirmation exercise, an intervention that CU researchers designed to mitigate stereotype threat, discussed in Chapter 3. The intervention consists of a brief writing exercise administered at the beginning of the course and again before the first exam. Students are asked to choose from a list of a dozen personal values the two or three values that are most important to them, then to write about why these values are most important, focusing on their feelings. The list of values includes relationships with friends and family, creativity, independence, and sense of humor. The justification given to students for doing the exercise was that it will give them practice communicating their ideas, and it will be important for them to be able to communicate their ideas about physics in the course.

The main topic for the second TIP class is a general introduction to problem solving. We begin by recognizing the wide variety of problems that students may encounter: practical everyday problems, puzzles, math and physics problems, research problems, and health diagnostic problems. For specifics, see the PowerPoints in Appendix 3. After considering various examples, we arrive at the Polya four step approach,[42] which can be applied to any kind of problem: (1) formulate the question; (2) plan a solution; (3) execute the plan; (4) review the solution. Application to practical problems helps students appreciate that planning the solution to a physics problem should engage their full creativity, as practical problems often do. Most importantly, the Polya approach gives students something to begin doing when they have no idea about how to solve a physics problem. Many students need to learn to become comfortable with their initial bewilderment at encountering a challenging problem.

I finish the second class by administering the Lawson test, again explaining that the purpose of this test is to find out what they already know.

The First Lab Meeting: Assessment and Games

During the first week of class, the lab period is devoted to completing assessment, by giving an attitudes survey, and then introducing students to some skill building games. I begin with Feuerstein type dot exercises that were described in Chapter 1. Examples are provided in Chapter 8. I hand out a packet of these exercises, but I ask students to complete only those on the first page during class and turn them in. These first dot exercises are relatively easy, and most students are able to complete them in a few minutes with 100% accuracy. However, there are always some students who make a significant number of mistakes. This is a clear sign that these students have impulsively connected the dots, for example creating a triangle out of 3 dots that are closest, even though that triangle has a much different shape than the one they are supposed to identify. This impulsiveness and lack of attention to detail, if not corrected, will make it very difficult for them to become good problem solvers. The dot exercises are very easy to grade; I return them next class, and I recommend that anyone who did not get 100% should continue to work on the dot exercises contained in the packet to improve their attention to detail. The grade is for diagnostic purposes only and does not contribute to a student's course grade.

Next I give a brief introduction to Sudokus, which many students are already familiar with and quite often enjoy. Like the dot exercises, Sudokus instill in the students an appreciation of the need to restrain impulsiveness, to be attentive to detail and to make changes when they make mistakes are. Such planning, monitoring and adjusting comprises key self-regulation skills in problem solving. Students are encouraged to continue to play Sudokus, and they are referred to Sudoweb.com, a website that provides Sudokus ranked according to level of difficulty.

The final game is "Kickbox," an ingenious computer game created by the MIND Research Institute.[43] A player progresses through a series of 60 games, divided into 10 levels. The games gradually increase in complexity and skill required (Fig. 5.2). To be successful, players must plan steps in advance. The higher levels make heavy demands on working memory. Some Kickbox games are provided on the TIP website (ThinkingInPhysics.com).

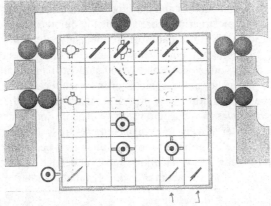

Fig. 5.2 A good student solution to a level 7 MIND game. The sketch shows how pieces on the grid should be positioned so that when a laser beam is emitted by the laser in the lower left corner, the beam can be reflected by the reflectors and strike the appropriate secondary lasers so that laser beams will knock all of the balls out of the path around the perimeter of the grid. This enables a penguin to travel around the perimeter. Planning is encouraged and guessing discouraged because if you fail twice on any game, you must go back to the first game in that level and work your way back to the game.

After this introduction, students can access the MIND games from their own computers to work their way through the various levels. Most students enjoy the games. As an added incentive, students can earn 10 extra credit points if they complete all games through level 8 and if they also take a printed quiz and show how to do two of the games selected from computer game screens. Completing level 8 is challenging and requires hours of work. Most students work on the games, and one third to one half of them successfully complete level 8 and the quiz.

Interviews
I interview each of my students for 15 minutes during the first week of class. I explain that the purpose of the interview is to get to know each of them and to make them more comfortable coming to see me for help with physics when they might need it. The interviews follow testing with the FCI, Lawson's test, an attitude survey, and their submission of their first thinking journal entry. I review students' first TJ entries along with assessment results before interviewing them. So by the time I meet with students, I already know a lot about them, their activities, their professional aspirations, their strengths and weaknesses, likes and dislikes – some loved their high school physics course, others hated it. Many are

concerned that they might not do well in my course either because they did not take high school physics, did not have a good course, or did not do well in it. I assure them that, based on the assessment I did of the class's initial knowledge of physics concepts (the FCI given on the first day of class), almost nobody in the class has a strong knowledge of these concepts and therefore they are not at a disadvantage. The class average FCI preinstruction score is typically below 30% and very few students score over 50%. The information about themselves and their goals that they shared in their first thinking journal entry makes the interviews easy for me and for them, and actually quite enjoyable. Many students have interesting stories – some have musical or athletic talents or interesting career aspirations. I discovered that one of my students is a professional athlete – on the U.S. national cricket team. Another loves to dance and wants to become known as the "dancing doctor." Even though interviewing sometimes requires up to 15 hours during the first week of class, it is well worth the time. Establishing a personal relationship with my students during the first week allows me to understand who they are and for them to feel more comfortable about the course and about talking to me.

6
TIP Course Structure

"Thinking is a bit uncomfortable, but you'll get used to it. A matter of time and practice." Lloyd Alexander

The regular sequence of instructional activities begins with the second week of class. For each topic in physics, the sequence is: reading the text, concepts class, homework problems, problem session, lab, quiz or test. Both concepts classes and problem sessions meet for 75 minutes; labs meet for 110 minutes. The class schedule in Appendix 4 shows the specific assignments for reading, homework problems, and labs throughout the first semester. TIP materials for each of these activities are provided in chapters 9–13. Additional materials are provided on the TIP website: ThinkingInPhysics.com.

Reading the Text; Online Reading Tests
Reading assignments cover concepts to be discussed in the next concepts class. Students are encouraged to read the text for a good understanding of the concepts, but not to spend much time on the worked examples in the text in this initial reading. Online reading tests provide students with an opportunity to take their first test on each new topic in an environment in which anxiety is minimized. The tests consist of five multiple choice questions and may be taken anytime up to 30 minutes before the concepts class. Using the Blackboard system, students are given three chances to answer the questions. I tell students that when their first try results in a score of less than 5, which is common, they should go back to the text and carefully read the material again. This proves to be a very effective way both to get students to read the book before lecture and to have them experience a physics test with very little anxiety. Most students score a perfect 5 by their third try. For those who don't, it is often because they are rushed for time and take their three tries in rapid succession, without going back to the book. Students' cumulative score on all reading tests counts for less than 10% of their grade, enough to motivate participation, but not enough to permit passing the course by using a surrogate test taker.

Seat Assignments
Based on the results of assessment and student interviews, I devise a seating chart that I present to students at the beginning of the second week of class, before the

first concepts class. Whenever possible I pair students of the same gender, though opposite sex pairs often occupy the same table. The assignment of partners is based on my intuitive guess about which students are likely to work well together. I try to avoid creating a pair in which the reasoning levels of the partners are too different, for example pairing a student who scores 90% on the Lawson test with one who scores 50%. I also try to avoid creating a pair or a table that has either much higher or much lower Lawson scores than most of the class. Because students do not choose their own seats, chances are low that best friends will be partners – an arrangement that could create distractions. A layout of the classroom with names and photos of students at their assigned seats helps me learn names quickly. A seating chart layout is shown in Appendix 5.

Concepts Class
My teaching is based on my belief that students should develop qualitative, conceptual understanding before beginning to solve computationally difficult problems. The first class meeting on each new chapter is a "Concepts" class, devoted to developing a purely conceptual understanding of physical principles in a Socratic style similar to Peer Instruction; the second class is devoted to problem solving. Concepts classes consist of a series of mini-lectures, punctuated by multiple-choice questions requiring a good conceptual understanding of the principles just covered.

Whenever possible, I try to make clicker questions interesting applications of the physics. For example, following a discussion of electric fields, I ask my students to imagine that they are hiking in the mountains during a thunderstorm and to choose the safest action from several options – life or death consequences depending on an understanding of electricity. (See Chapter 10 for details of this question and many others.)

When a question is posed, students usually have about two minutes to respond using a clicker that is attached by a chain to each seat. This arrangement allows for each student in each class to have a clicker number, so that I can keep track of individual responses. I use the iClicker system. The distribution of answers is displayed as a histogram on the projector screen. The display also shows who has responded, but not their answer. Thus I can immediately see whether anyone has not responded. Sometimes it takes a little prodding, but I generally get 100% participation. Students are not graded on their answers to clicker questions, but they are given a small number of points for participating – about 2% of their grade.

When the great majority of students get the correct answer, I call on someone to explain. After the explanation, I ask if there are any questions. More interesting and more typical is when few students have the right answers, often with a wide mix of answers. I then ask students to discuss the question with partners and try to persuade them. This usually takes several minutes, at the end of which students vote again, usually with a very different distribution of answers. When the questions are not very difficult, this second round often results in a large majority having the correct answer. When this doesn't happen, I give a hint and encourage another round of discussion. In most cases, at least by the third try almost all students have arrived at the correct answer – without my having given it to them. Occasionally, many students get the wrong answer even after the third try. I then ask someone in the class to defend the most popular wrong answer and another to defend the right answer, without yet indicating which answer is correct. If necessary, I do eventually explain the correct answer. The discussion can take a significant amount of time, but it ensures that students are strongly engaged with the concepts. On the first day of class and in the Guide to Learning Physics, I stress the importance and effectiveness of this way of learning concepts. Almost all students accept this approach.

I encourage students to continue to engage with these concepts after class by reviewing the PowerPoints and clicker questions, which I post on Blackboard, and by rereading the text and trying end of chapter concept questions.

Homework Problems

Before they start to solve assigned homework problems, I encourage students to read the text again, this time paying close attention to worked examples in the text. I suggest that they initially cover up the solution and try to provide their own solution, and if they get stuck, which is likely, to uncover just enough of the solution to get started again. Working their way through the examples prepares them to tackle end of chapter problems. I point out that some of the problems may be simple exercises that they immediately know how to solve. In other cases, the problems may be true problems, meaning that initially they don't know how to solve them. Throughout the semester, whenever they are solving a problem, I stress the importance of following the Polya steps of problem solving. I also point out the almost universal tendency to want to rush through problem solving. Everyone wants to get an answer quickly, but slowing down and being very careful in the beginning of a problem often makes problem solving more efficient and quite often even faster. Answers to most of the assigned problems are in the back of the book. If they are stuck, there are many resources for help, but I urge students to not give up too quickly. The solutions manual provides solutions to some of the assigned problems. These solutions should be used sparingly, in the same way suggested for the worked examples in the text.

Pencasts

I provide recorded "pencast" solutions to selected problems. These are QuickTime movies on Blackboard. The recorded audio and pen strokes can be started and stopped. Examples are provided on the Thinking in Physics website. A pencast provides a model of all the elements of careful problems solving – formulating, planning, execution, and review. I urge students not to view the movie passively. Instead, I ask them to provide each step of problem solving before viewing the recorded problem solving step. After viewing each step, they are asked to reflect on their solution and whether their step included essential elements that were present in the pencast solution. For example, did their formulation include an adequate drawing? This element of self regulation tied to a specific problem is helpful in developing a more systematic approach to problem solving.

Problem Sessions
Problem sessions provide an opportunity to go through assigned homework problems that students have been working on. Most students come to class having made some progress on the homework, but with much work to be done. I ask them to begin by working with their partners and others at their table to compare their solutions. While they do this, I circulate around the classroom to get a sense of how far each student has progressed. Homework is not collected, so there is no grade associated directly with this part of the course. Weekly quizzes provide incentive to keep up with the homework. Some students have already solved every assigned problem by the beginning of the problem session, while others have barely started solving problems. In order for everyone to work together on a few problems, I add a few textbook problems, not previously assigned, to be worked on during and after the problem session. Typically, the level of these problems is challenging – true problems, not exercises, for everyone in the class. Students work first individually, then in groups, on each of the Polya steps of problem solving. After most have completed a step, we go over that step as a class. To encourage the use of these steps, I have developed problem solving worksheets that provide a template for their solutions. For each problem there are two worksheets. Students use Worksheet A to formulate the problem and to begin the plan for a solution by identifying the general type of problem they are working on. Then they use the version of Worksheet B that is appropriate for that type of problem – kinematics, dynamics, energy, etc. For a student who views problems in terms of their superficial features and believes that there are an enormous variety of problem types, it can be illuminating to see that in a fundamental sense the number of problem types is very small. I emphasize the importance of drawing a sketch on Worksheet A and of expressing what they are trying to find, in terms of symbols (e.g. find x when $v_x = 0$). Chapter 11 provides some example worksheets. More are available on ThinkingInPhysics.com.

Some problems are estimation problems, in which no data are provided. Students must estimate all numerical values used in their solutions. An example taken from a test (Ch. 13) is estimating the force exerted by a goalie's foot on a soccer ball when it is kicked the length of a field. Such problems encourage students to tap their own resources, their understanding of basic concepts, rather than to hunt for equations involving variables whose values are supplied.

Working problems at the blackboard can have some value, but I don't like to spend too much class time on this. Working one-on-one with students during office hours, my help can be more focused on individual needs. The availability of pencasts answers the students desire to see more worked problems without using valuable class time. These pencasts can be played back as many times as necessary, making them in some ways more valuable than blackboard problem solving.

Office Hours
I encourage students to see me for one-on-one help during office hours or by appointment. This can be especially helpful in diagnosing individual difficulties students encounter – reading the text, understanding difficult concepts, solving

problems, or sometimes addressing special needs. By examining students' course schedules online, I am able to choose office hours that do not conflict with other classes for nearly all students in a class. Nevertheless, some students have many other commitments – athletics, jobs, internships, clubs – making it very difficult to attend office hours, and so I try to make special appointments to meet with them.

Labs

The 20 to 30 students who are in a section of my TIP course and meet in the TIP classroom for concepts classes and problem sessions are divided into two groups and meet in small adjacent lab rooms, each with six tables. Again, seating is assigned, with two or three students per table, depending on class size. I am aided by two teaching assistants, one in each room. The TAs are typically my former students who love the work. I meet with my TAs for training each week for one hour. They go through the week's labs in detail and I model for them the kind of assistance I want them to provide for students. I emphasize the importance of resisting the common impulse of wanting to just give answers. Because the ratio of students to instructors is only between about 7 to 1 to 10 to 1, the TAs and I are able to provide very individualized guidance.

Labs are focused on deepening students' understanding of physics concepts. I find the labs to be more productive when students believe that labs give them the opportunity to better understand concepts. To encourage that point of view, I eliminate concern about grades in lab. Full credit, about 14% of total points, is given for lab attendance, good faith effort, and participation in the lab.

I try to make the labs fun. The most popular lab is one related to Newton's second law in which students pull each other on carts, sheets of plywood that roll on low friction, Rollerblade wheels. One student pulls the cart on which a second student is seated, while the student who is pulling maintains a constant force as the cart moves about 15 m down a hall. The cart is pulled with a rope and spring scale, so that the student who is pulling can monitor the force during the motion. Students are often surprised to find that they are initially not able to maintain a constant tension, say 20 N (significantly greater than the friction force), because they do not realize that they must accelerate to do so. The riding student uses a metronome and marks the floor at one-second intervals so that the distance traveled as a function of time can be measured. This experiment serves to develop both students' understanding of Newton's second law and also their facility with handling multiple variables and their relationships. See Chapter 12 for details on this and other labs.

Lab instructions are integrated with questions that guide students to perform measurements and to analyze data. Students do not turn in lab reports. They are told to hold on to the lab instructions and questions, and to review them in preparation for tests. In addition to actual experiments, many labs also have worksheet questions that utilize computer simulations of experiments, often from the PhET website.[44] Excellent simulations on the PhET website can be reviewed after lab.

Tutoring Selected Students

The assessment that I do in the first week enables me to identify those students who are likely to have the greatest difficulty with the course, 4 to 6 students for each lecture section of about 30 students. I assign 2 or 3 students each to my two TAs for that section. I invite these students to receive an hour or two of tutoring each week from the TAs. The students almost always accept. I check with the TA tutors on their students' progress at weekly lab prep meetings.

Lab Concept Quizzes

A lab concept quiz is provided for each lab. No grade is attached to the quiz, which is taken at the end of the lab and then discussed by the class. The discussion is led by the TA or by me. The quiz helps students gauge their understanding of the lab. I encourage students to review these quizzes in preparation for quizzes and tests.

Quizzes and Tests

Quizzes are given weekly on topics covered the preceding week. Each 15 minute quiz is worth 10 points, about 1.4% of the total points in the course. Two full period (75 minute) midterm tests are each worth 100 points, about 14% of the total. The two hour final exam is worth 200 points or about 28% of the total. Points on quizzes and tests are about equally divided between concept questions and problems. Quizzes are open book; tests are closed book, but students are allowed a formula page, two pages for the final. See Chapter 13 for sample quizzes and tests.

Frequent quizzes encourage students to keep up with the material and to make a serious effort on homework. Quiz questions and problems are pretty basic, and class averages on quizzes are usually about 70%. Some questions and problems on tests are more challenging. Class averages are usually 55% to 65%, with a wide range of individual scores, mostly in the range of 40% to 80%. Because my students are not accustomed to such low test averages, in the Learning Guide I am careful to explain the reasons for the challenging level of the tests. I stress that physics is hard and any test that measures their understanding of physics must do so by probing for real understanding, not by simply asking them to write solutions to problem types that they are so familiar with that they are merely exercises. At least some of the problems on each test are true problems, requiring them to think. Concept questions too are thought provoking. I consider a score of 70% on these tests a good score, indicating good understanding of physics concepts and good problem solving; 80% is an excellent test score. Though tests are harder and test scores are therefore lower in my class than in most of their other science classes, the distribution of grades is similar to that of other science classes, at times even higher. In recent Thinking in Physics classes the average grade has typically been B or B-. Students do not compete with others for a grade in my classes. There is no limit to the number of A's that can be earned.

I urge students to regard tests as not only a measure of their understanding, but also as an opportunity for further learning. I encourage them to review their graded test, correct anything that was wrong on a new copy of the test, reflect on what it was about their thinking that led to mistakes on specific questions and problems, and submit corrections and reflections one week after the graded test is returned. Students earn 25% of any deducted test points for which they supply the correct solution and a reflection. I encourage students to work as independently as possible on the corrections, but to feel free to discuss with others and to see me for help if needed.

Thinking Journals

In addition to the first thinking journal entry assigned at the first class meeting, I assign several other TJ entries throughout the semester. Prompts include asking whether their thinking in physics has changed since the beginning of the course, and asking them to comment on quotes related to the nature of learning by students or others. Here are some example TJ comments:

> "So far, the concepts that I have learned in class are straight forward – until you add word problems! I don't know what it is about word problems, but there always seems to be something that I miss that is a big part of the problem and dictates an additional step or calculation. From the homework that I have been doing, I feel I just need to get into the hang of these questions. I need to start getting into the feel of them and feeling comfortable with the way they are worded and what it is they are asking."

> "I've decided my motto for this class will be: 'slowly but surely.' I need time to digest and really think about the material I've been learning."

> "I got a little bit excited because the concepts of velocity, acceleration, relative motion, just clicked. I was able to tie the graphs of position, velocity, and acceleration together – it is just all about the DERIVATIVES. It was seriously a great moment for me to just have that click and see the problems in a completely different light – and see that they weren't that impossible. Relative motion is a little bit fun for me because it's just seeing things in different perspectives and applying the concepts."

Group Project

Students receive 25 points (about 3.6% of total points) for participating in a group project. The project teams four students seated at the same classroom table, making discussion of projects convenient and the total number of projects to be presented small. Each group creates a five minute YouTube video of their project to be shared with the class at the end of the semester. The purpose of the project is to help develop interest in physics. The idea is to find something that the group is already interested in and for which they can apply the physics they learn during the semester. Viewing the group projects at the end of the semester requires very little class time for eight groups in a class of 30.

Many years ago I had assigned projects, but had given it up in recent years because it consumed too much of my time directing students' efforts to correctly explain the physics in their project. I now leave students completely on their own. Students get full credit (about 3% of their grade) for a good faith effort producing and presenting to the class a short video without my involvement. I no longer see the projects before they are presented to the class, and so I do not spend time on the project myself. The Web provides the opportunity for exploring the physics of virtually anything you can imagine: dance, surfing, bungee jumping, pool, flycasting, boomerangs, motocross racing, and of course any kind of sport. The result is that many groups create videos that are imaginative, often humorous. Even though the physics in some projects is not completely understood, I resist my impulse to help because the project is all about building students' interest in physics by working on something they are already interested in. I stress that, though many of them are not that interested in physics, surely they are interested in *something*, and since physics applies to just about everything, the project is an opportunity to develop their interest in physics by applying it to something they are interested in.

Students tell me that they appreciate seeing applications of physics in other groups' presentations. In an anonymous poll at the end of a recent class, 90% of the students indicated that they thought I should require projects in future classes.

Last Class Meetings; Post Instruction Assessment
The last class meeting and last lab meeting of the semester are devoted to showing project videos (about 45 minutes), post-instruction assessment using the FCI, Lawson's test, a quiz based on the Kickbox games, and an attitude survey (about 90 minutes), and a student evaluation of teaching (about 10 minutes). To encourage maximum participation and serious effort on all the assessment at the end of the course, I offer extra credit. Students are told at the beginning of the course that if they complete the first 8 levels of the Kickbox game and then successfully complete a quiz based on the higher levels of that game, they will earn 10 points of extra credit. Nearly all students begin the games and most thoroughly enjoy them. But, because the lower levels are easy and the higher levels are quite hard, quite a few students underestimate the time and effort required to complete all 8 levels, and only about one third to one half of the class completes level 8 and takes the quiz. Nearly all who take the quiz receive 10 points of extra credit.

When I give the FCI on the first day of the course, I tell students that it is a test to find out what they already know about physics, and that, although the test does not count toward their grade, it is important for them to take it seriously. I describe the Lawson test in a similar way when I give it on the second day of the course. A few days before the postinstruction FCI is given, I tell my students that they should begin reviewing the basic concepts they have learned in the course, and to encourage that, I offer an opportunity to earn extra credit based on their perfomance on a test of those concepts. I tell them that they will earn one extra credit point for each correct answer, but I am intentionally vague about the number of questions, telling them there will be about 20 or 30 questions. Students do not know that they are taking the same test they took on the first day of class until they are taking it. I do not announce the Lawson Test in advance. Rather, as they are nearing completion of the FCI, I offer the Lawson test as another extra credit opportunity. Students earn one point for each correct answer, minus 12 points, for a maximum of 12 extra credit points. The maximum number of extra credit points that a student can earn for the Kickbox quiz, the FCI, and the Lawson test is 52 points, or about 7% of the total points in the course. The average number of extra credit points earned is 30, or about 4% of the total points in the course. I believe that it is important to offer extra credit so that students will take the tests as seriously as they do on the first class day. If the tests were optional with no credit provided, a significant number would choose not to take them. If only participation points were offered, some would take the tests, but not make a serious effort on them. I make sure that all copies of the FCI and the Lawson test are accounted for and collected, and I never use the name of either test.

Student Evaluations of Teaching
At the end of the last class meeting, I ask a student to administer the anonymous, university sponsored, student evaluation of teaching, and I leave the room while students complete the forms. Results of these evaluations, both individual forms and averages, are given to me after grades are submitted for the course. The evaluation form uses a 5-point scale (5 = very good, 4 = good, 3 = fair , 2 = poor, and 1 = very poor) to rate the instructor in eight categories, including addressing learning outcomes, availability for discussion, feedback that improves learning, and overall effectiveness. Students are also invited to write comments related to each of the categories. Two more optional, instructor generated questions give an instructor the opportunity to ask for students' opinions about specific course elements. Over the years, average scores on my student evaluations have been about average for the university. These scores did not change appreciably when I began using more interactive methods, or even when I began the TIP curriculum, even though students learned much more than they had previously. However, with the most recent changes to TIP, which included the Learning Guide, individual student interviews in the first week, and group projects, student evaluations were significantly higher. For the three years before these changes (2009 to 2012), the average over all eight questions on my students' evaluations in the first semester course (PHYS 253) averaged 4.1. In the two years after the change (2012 and 2013), the eight question average in PHYS 253 averaged 4.5. Even students' average rating of my availability improved from 4.4 to 4.7, though my actual availability was unchanged.

Student evaluations are useful in providing students' opinions about the course. Student comments are particularly valuable in this regard. For example, I know from these comments that students like pencasts and believe that they are helpful.

These comments also show that students who give the lower numerical ratings would prefer for me to work more example problems and give tests that are easier, with problems that are more similar to homework problems. In other courses students are often used to seeing only very specific problem types that they have been trained to solve. Some of these students feel that my problems are unfair. To try to counter that attitude, one of my thinking journal prompts asks whether my test problems are fair. I then ask whether, if they were physicians or physical therapists, it would be fair for a patient to come to them with a problem that they had not been trained to solve. I try to make the point that learning to become flexible problem solvers, willing to struggle with problems that seem different from anything they have seen before, will have some value in their lives beyond my course.

Despite the value of student evaluations, one should never use them as a measure of teaching effectiveness.

Carl Wieman has suggested that colleges and universities begin using a teaching practices survey[45] that he and Sarah Gilbert developed in order to encourage the improvement of science teaching. He proposes that, instead of basing rewards on the results of student evaluations, instructors be rewarded on the extent to which their self-reported (and easily verified) teaching practices align with research-based best teaching practices. Wieman refers to the recent meta-analysis[46] of the validity of teaching evaluations by Clayson, who wrote "Objective measures of learning are unrelated to the SET (student evaluations of teaching). However, the students' satisfaction with, or perception of, learning is related to the evaluations they give." Wieman[47] writes:

> These shortcomings of student evaluations were confirmed at the Science Education Initiatives at the Universities of Colorado and British Columbia which changed the teaching of about 150 courses. The student course evaluations for the faculty involved are largely the same before and after they have transformed their teaching practices, although quite different teaching methods are being used that often produce measurably greater amounts of learning.

We have presented a survey of teaching practices that provides a rich and detailed picture of what practices are used in a course. We also have presented a scoring rubric that gives a quantitative measure of the extent of use of teaching practices that research has shown are consistently effective in achieving improved student learning. This is a more practical and accurate measurement of teaching quality than any existing alternative. In addition to being a useful tool for self-reflection of teaching, the survey provides a straightforward and objective way to compare the quality of teaching across the courses and faculty members within a department, and to make comparisons of the quality of teaching practices used across STEM departments both within and between institutions. We believe this will be a valuable tool for evaluating and improving undergraduate STEM teaching.

7

TIP at Chicago State University
by Mel Sabella

"A person with Ubuntu is open and available to others, affirming of others, does not feel threatened that others are able and good, for he or she has a proper self-assurance that comes from knowing that he or she belongs in a greater whole and is diminished when others are humiliated or diminished..."   Desmond Tutu

Since 2004, the physics program at Chicago State University (CSU) has been using modern, research-based teaching techniques and materials, and conducting research into the effectiveness of this curriculum. Grant support from the National Science Foundation and the American Physical Society has helped us develop and incorporate a set of instructional tools that serve the needs of underrepresented students, who bring a unique set of resources and needs to the introductory physics course. The quote above from Desmond Tutu describes a set of community resources that our students bring to the science classroom. We find our students rely heavily on the "we" in the classroom and support each other through dialogue. This resource aligns well with the innovative pedagogies we utilize in our classes.

    CSU is approximately 80% African American, and serves students from the surrounding community, the South Side of Chicago. As a result of our course revisions, we have seen improvements in conceptual understanding and student attitudes toward physics[48], including improved gains on diagnostics such as the FCI. Before the revised curriculum was introduced, normalized FCI gains averaged 0.2. Shortly after the revisions, average gains improved to about 0.3, which is on the low end for IE courses. According to Vince Coletta, these scores are in line with IE normalized gains for students whose scores on standardized admission exams are similar to ours. (Our students' average ACT score is about 18, which is considered roughly equivalent to an SAT math + verbal score of 875). The mean for such students at LMU and ELHS was 0.25, according to Fig. 2.1b, which shows pre-TIP data. We analyzed gender differences during one semester of instruction and found that females had higher gains than males, supporting Coletta's conjecture of a reverse gender gap for low SAT students.[37]

Although our improved gains are encouraging, especially when compared with others with similar ACT or SAT scores, they are still modest. Most students' post instruction FCI scores are below 50%. This is a concern. Yet open-ended questions often show that students have a better understanding of physics concepts than diagnostic test results indicate. This suggests that our students' metacognitive skills may be limiting their performance on the FCI. One way to look at poor results on the FCI is that students simply do not understand the underlying concepts. Instead, we are suggesting that these results mean that our students are unable to trigger the appropriate concepts when responding to multiple-choice diagnostics, even though they have a basic understanding of the concepts, and can often arrive at the correct answer when verbally responding to the same questions they answered incorrectly on the multiple-choice test. An example that illustrates this comes from interview data we collected as students responded to questions from the Force and Motion Concept Evaluation (FMCE).[49,50] When we administered these questions in class, students would typically perform quite poorly on the questions. In a follow-up, we conducted one-on-one interviews, examining in detail students' responses to FMCE questions, such as the following one, in which a sled is moving on frictionless ice:

> Choose the one force which would keep the sled moving.
> A) The force is toward the right and is increasing in strength (magnitude).
> B) The force is toward the right and is of constant strength (magnitude).
> C) The force is toward the right and is decreasing in strength (magnitude).
> D) No applied force is needed.
> E) The force is toward the left and is decreasing in strength (magnitude).
> F) The force is toward the left and is of constant strength (magnitude).
> G) The force is toward the left and is increasing in strength (magnitude).

In an interview, a student who had answered this question incorrectly on the test was asked, "What force would keep the sled moving to the right at a constant velocity?" She responded:

> So, if you're pushing the sled you're causing it to move and you let go, let go, and you stop pushing it then the sled would continue to move ...
> ... but the motion would just decrease. The motion would decrease in velocity.
> ... No, let me change that answer ... Okay, this is my final answer.
> ... If you're applying a force ... and you stop pushing it, Newton's first law says that an object in motion stays in motion so it's just going to keep on moving.

Her response shows that she understands Newton's laws, but has trouble triggering this formal knowledge, and she struggles with her intuition and everyday experiences, and how these fit with the formal knowledge. Questions from the FCI or FMCE tend to activate in our students an impulsive, intuitive answer, rather than a formal answer based on course content. Because of this, we began to look at other instructional materials that focused on thinking skills, such as controlling impulsiveness. After discussions with Coletta and Phillips, we decided to offer a version of the TIP curriculum to supplement our introductory physics course. To me, this seemed like the missing piece.

The importance of pilot testing instructional materials with different populations is extremely important. Too often, an implicit assumption is made that a curriculum created at one institution with a particular kind of population is transferable to any population. Students at CSU are quite different from students at LMU and many other schools, in terms of culture, family responsibilities, socio-economic status, and average ACT scores. Most of our students have grown up close to our school and have attended high school on the South Side of Chicago. About 90% of our students receive financial aid and our students tend to be older than the typical college student. Our school is about 70% female and many of our students have full time jobs and families. We also find that our students have an excellent set of resources that other students may not have. They value the tools and methods of reform-based physics instruction almost immediately, they engage in peer discussion very easily, welcome questions from the instructor, and have a strong sense of community and peer support.[51] The opportunity to test some of the TIP materials with our students and discuss the implementation and student feedback with Coletta was extremely useful.

In 2009 we offered a 1 credit course called Practical Skills for Success in Science. The course was offered as an elective for students enrolled in Physics I, a calculus based course. The class met for one hour each week. Ten of the fifteen students in this course were also enrolled in the Practical Skills Course, which used a selection of TIP materials and some CSU created activities to help students develop specific skills that we hoped would improve their performance in the introductory physics course. Components of the course included:

- Estimation problems
- Sudokus
- Kickbox – from the MIND Research Institute
- Problem solving (physics and beyond, Polya's four steps)
- Connecting the Dots
- Thinking journals
- Articulation of strategies for different tasks
- Experiments (Control of Variables, Influence and Determine)

The course began with an estimation problem that used the PhET Simulation "Estimation."[48] This is a game in which students estimate the number of circles that fit in a given area or the number of short line segments that make up a long line segment, etc. Once students played the game a few times, we discussed the types of strategies they developed. They were then asked to estimate the number of tennis balls that would fit into our classroom. These initial activities align with Coletta's goals of helping students overcome impulsiveness, pay attention to detail, and develop personal strategies. At this point students began to record their strategies in thinking journals. Students were then asked to think about the types of strategies they use when solving problems. Once students had written their problem solving strategies, we compared them to Polya's approach, as used in TIP.

As the course progressed, we utilized more of the TIP materials, such as the Connect the Dot puzzles and Kickbox computer games from the MIND Research Institute. Kickbox was especially enjoyable for the students as they engaged in friendly competition with each other to see who could get to the highest level. Many students worked on Kickbox outside of class, despite receiving no credit for it. Students enjoyed the challenge of the game, and solving challenging levels instilled a sense of confidence and self esteem.

Students showed some improvement in Lawson scores, with an average normalized gain of 0.2. There was no improvement in FCI gains. However, we used only a sample of some of the earlier TIP activities and the intervention was only one hour per week. It was clear to us that engaging in this type of activity for one hour a week may not be enough to see substantial FCI gains.

Students were quite positive about the TIP activities, and they saw a connection between the skills they were developing in the Practical Skills course and other courses they were taking. One student commented:

> In my science and society class we were studying population patterns in Indiana and we had to test the numbers we got to see what kind of information we could obtain. Because of my physics 1010 class, I took my time to analyze the data and looked at everything that was given without jumping to conclusions or assumptions. Those skills also helped me a lot with my test taking skills. I used all the strategies I learned to answer test questions. I read the questions, took note of what was given, drew a picture to represent the problem, did not rush and did the problem. In the questions that I did not know how to answer I simply did something.

Another student remarked:

> I learned to take my time, to look at what is given, to expect the unexpected, not to give up, to draw a picture to go along with the problem, plan a strategy, and to make sure I understand the problem. I learned to develop these techniques with the other problems we did in class.

Another student stated that the course:

> taught me a lot of skills and different ways of looking at problems. Since day one Physics 1010 has helped me expand the way I think. The different strategy games and puzzles, such as Kick box and Sudoku, have helped me with other classes. Throughout this course one of the main strategies I learned was to be patient when solving a problem.

Comments like these show that our students see a great deal of value in the TIP materials and they see a connection between the skills they develop in this type of course and their other courses. Fitting TIP activities into already busy student schedules is a challenge, but one that seems worthwhile.

Part III
TIP Materials

8 Connecting the Dots

9 Reading Tests

10 Clicker Questions

11 Problem Solving Worksheets

12 Labs and Lab Concept Quizzes

13 Quizzes and Tests

8
Connecting the Dots

"Thinking is learning all over again how to see, directing one's consciousness."
Albert Camus

Connecting the Dots

Connecting the Dots

9

Reading Tests

"Think before you speak. Read before you think."   *Fran Lebowitz*

Each assignment indicates both the topics covered and the specific sections of my textbook (*Physics Fundamentals*) assigned for reading. The tests should be usable with any introductory text.

First Semester

Reading 1: Measurement, units, vectors, speed, velocity (Meas. & Units and Ch. 1)
Reading Test 1
1 In the equation $y = \frac{1}{2} at^2$, y is in units of m and t is in units of s. What are the units of a? A) m/s; B) (m/s)^2; C) m/s^2; D) m^2/s^2

2 How many significant figures are in the number 12.50?
    A) 1; B) 2; C) 3; D) 4

3 You can estimate a car's instantaneous speed from: A) how long the car has been traveling; B) how long it takes for the wheels to complete one turn; C) the road's speed limit; D) the pressure you exert on the gas pedal.

4 Vector addition of vectors means finding a resultant by:
    A) connecting vectors tail to tip and drawing the resultant directed from the tip of the last to the tail of the first; B) connecting vectors tail to tip and drawing the resultant directed from tail of the first to tip of the last; C) adding the lengths of the vectors.

5 The instantaneous velocity of a car means: A) how fast the car is traveling at some instant; B) how fast the car is traveling at some instant and the direction it has moved since beginning its motion; C) the total distance traveled by the car divided by the time; D) the readings of the speedometer and compass on the car's dashboard.

Reading 2: Acceleration, linear motion at constant acceleration, graphical analysis, relative motion (Sects. 2-1, 2-2, 2-4, 3-1, & 3-4).
Reading Test 2

1 Acceleration means: A) increase in speed; B) change in velocity; C) rate of change of velocity; D) rate of increase in speed.

2 The symbols for an object's speed, its velocity vector, and its x-component of velocity are: A) $s, v, v(x)$; B) $v, \mathbf{v}, v_x$; C) $v, \mathbf{v}, v(x)$; D) s, v, v_x.

3 When using the equation $x = x_0 + v_{x0}t + \frac{1}{2} a_x t^2$, the symbol x means:
A) how far the object has traveled at time t; B) the direction of the object's motion; C) the average distance of the object from the origin over time t; D) the position of the object on the x axis at time t.

4 For constant, positive acceleration, graphs of x vs. t and v_x vs. t will be:
A) straight lines for both x and v_x; B) a straight line for x and a line that curves concave upward for v_x; C) a line that curves concave upward for x and a straight line for v_x; D) lines that curve concave upward for both x and v_x.

5 Which of the following statements is true? A) Sometimes a body may appear to be moving, but it really isn't moving. It only appears that way because you are viewing it from the wrong place; B) All motion is relative to the observer; there is no absolute motion; C) If body X is moving relative to body Y, and body Z is moving relative to body Y, then body Z must be moving relative to body X; D) The earth's surface is always the best reference frame for describing the motion of any body.

Reading 3: Newton's laws of motion (sects. 4-1 to 4-5).
Reading Test 3

1 Mass is: A) a measure of a body's resistance to being accelerated; B) dependent on where a body is located; C) the force of gravity's pull on a body.

2 The mass of a body can be measured in: A) pounds or kilograms; B) kilograms or newtons; C) grams or kilograms; D) pounds or newtons.

3 An object that is initially moving will: A) keep moving only if force is applied to it; B) keep moving if no force is applied to it.

4 Newton's second law relates the acceleration of a body to: A) the displacement of the body; B) the velocity of the body; C) the resultant force acting on the body.

5 Newton's laws of motion are valid: A) in all reference frames; B) only in the earth's reference frame; C) only in inertial reference frames.

Reading 4: Force laws; applications of Newton's laws (sects. 4-6 to 4-8)
Reading Test 4
1 The gravitational force exerted on an object that is on or near the surface of the earth, commonly referred to as the weight of the object:
A) means the same as the body's mass in kilograms; B) is a force that can vary in magnitude by as much as 10% depending where on earth the object is located; C) is a force that acts on the object only if it is in contact with some part of the earth; D) is a force exerted on the object by the entire earth; E) is not equal to the force exerted by the object on the earth.

2 The gravitational force exerted on an object that is on or near the surface of the earth is: A) always in the same direction independent of where the object is on earth; B) in one direction at a given point on earth, but would be directed in the exact opposite direction if the object were moved to the opposite side of the earth.

3 The tension force in a flexible string, rope, or cable: A) can be either a push or a pull; B) can point in one direction or in the exact opposite direction, dependent on what you choose as the free body on which the tension force acts; C) is not always parallel to the string, rope, or cable; D) in a tug of war will be greater in the direction of the team that is winning than in the direction of the team that is losing.

4 In choosing an object as a "free body" for which you draw a free-body diagram:
A) it is valid to choose any body; one free body is as good as another; B) it is valid to choose any body, but not all free bodies are useful for solving a particular problem; C) it is only valid to choose complete, whole bodies as a "free body"; never choose part of a whole body as a free body.

5 In drawing and analyzing the forces that act on a free body in a free-body diagram
A) you should include all of the forces exerted on the free body by other objects and also all of the forces exerted by the free body on other objects, and you can use any reference frame to analyze those forces; B) you should include all of the forces exerted on the free body by other objects, and you can use any reference frame to analyze those forces; C) you should include all of the forces exerted on the free body by other objects, and you can use any inertial reference frame to analyze those forces; D) you should include all of the forces exerted on the free body by other objects and also all of the forces exerted by the free body on other objects, and you can use any inertial reference frame to analyze those forces.

Reading 5: Projectiles (sects. 2-3 and 3-2)
Reading Test 5
1 When air resistance is negligible: A) all objects fall with the same constant acceleration of 10 m/s^2; B) an object falls with an acceleration that depends on the mass of the object; C) an object falls with an acceleration that depends on the height from which it falls; D) an object falls with an acceleration that is less at the beginning of the fall and greater near the end of the fall.

Reading Tests

2 An object is thrown vertically upward. The object initially goes up, then comes down. Air resistance is negligible. A) The object's acceleration while coming down is different from its acceleration while going up. B) The object's acceleration is the same throughout its motion. C) The object's acceleration is the same throughout its motion, except at its highest point where its acceleration is zero.

3 An object is given an initial velocity, directed at some angle above the horizontal. Air resistance is negligible. A) The object's horizontal and vertical components of velocity, v_x and v_y, are constant throughout the motion; B) The object's horizontal and vertical components of velocity, v_x and v_y, both decrease throughout the motion; C) The object's horizontal component of velocity v_x is constant throughout the motion, but its vertical component of velocity v_y decreases throughout the motion; D) The object's horizontal component of velocity v_x is constant throughout the motion, but its vertical component of velocity v_y decreases while the object is rising and then increases while the object is falling.

4 When air resistance is negligible, the trajectory of a projectile: A) is circular in shape; B) is parabolic in shape; C) is neither circular nor parabolic.

5 Two objects, A and B, are given the exact same initial velocity, directed at some angle above the horizontal. The mass of B is significantly greater than the mass of A. Air resistance is negligible. A) The gravitational forces acting on A and B are the same and their trajectories are also the same; B) The gravitational forces acting on A and B are the same but their trajectories are different; C) The gravitational forces acting on A and B are different and their trajectories are also different; D) The gravitational forces acting on A and B are different, but their trajectories are the same.

Reading 6: Friction (sect. 5-1)
Reading Test 6
1 A book is on a tabletop and may or may not be moving along the surface. The force that the horizontal surface of the table exerts on the book: A) is always horizontal; B) is always vertical; C) always has both a vertical and a horizontal component; D) may have both vertical and horizontal components, depending on other forces that may be acting on the book.

2 A friction force exerted by a table's surface on a book on the table:
A) always acts on the book whether or not the book slides along the surface;
B) acts only if the book slides along the surface; C) acts on the book if the book slides along the surface or if the book does not slide along the surface, but some other force is pushing the book in a direction along the surface.

3 The normal force exerted by a surface on an object in contact with the surface is always: A) vertical; B) horizontal; C) perpendicular to the surface.

4 The magnitude of the normal force exerted by a surface on an object of mass m is always: A) equal to mg; B) equal to $mg \cos(\theta)$, where θ is the angle between the surface and the horizontal; C) just big enough to keep the object from going through the surface.

5 An object is at rest on an inclined surface, and no forces act on it other than its weight and any force that the surface might exert. A) There is a friction force on the object, directed down the incline; B) There is a friction force on the object, directed up the incline; C) There is no friction force on the object.

Reading 7: Circular motion and center of mass (sects. 3-3, 5-2, & 5-3)
Reading Test 7
1 A body moving at a constant speed along a circular path experiences: A) no force; B) no net force but an inward force and an outward force that cancel; C) a net force directed inward toward the center of the circle; D) a net force directed outward from the center of the circle.

2 Can tension and/or friction forces produce centripetal acceleration?
A) yes for both; B) yes for tension, no for friction; C) no for both;
D) yes for friction, no for tension.

3 The center of mass of a solid metal sphere: A) is on the surface of the sphere; B) is at the center of the sphere; C) depends on the mass of the sphere.

4 If no external forces act on an object and its center of mass is initially at rest, the center of mass: A) moves at constant velocity; B) does not move; C) accelerates; D) moves only if the combination of internal forces is right.

5 If you toss a spinning baton into the air, its center of mass will follow a trajectory that is: A) a circular path; B) unpredictable because it depends on how fast the baton is spinning; C) unpredictable because it depends on the baton's weight; D) the same as that of a particle, a parabolic arc.

Reading 8: Gravitation and momentum (sects. 6-1, 6-2, 8-1, & 8-2)
Reading Test 8
1 The law of universal gravitation describes: A) the force that acts on a body only when it is near the Earth or some other planet; B) the kind of force that acts on any body only when it is at the exact center of the universe; C) the repulsion between any two bodies; D) the attraction between any two bodies.

2 The universal gravitational constant G is: A) different on the moon than on earth; B) equal to g (9.8 m/s^2) at the surface of the earth; C) the same everywhere in the universe.

3 The total momentum of an isolated system: A) always increases; B) always decreases; C) neither increases nor decreases; it is constant; D) is independent of reference frame.

4 What physical characteristic(s) of earth determine(s) the value of gravitational acceleration on the earth's surface? A) density; B) distance from sun; C) mass; D) radius; E) mass and radius.

5 When a shotgun is fired and recoils, compared to the momentum of the shot that is fired, the momentum of the recoiling gun is: A) more; B) less; C) the same.

Reading 9: Energy, part I (sects. 7-1 & 7-2)
Reading Test 9
1 The total energy of an isolated system is: A) always conserved; B) never conserved; C) sometimes conserved.

2 The work done on a body moving in a straight line by a constant force of magnitude F, directed at an angle θ relative to the direction of motion, is: A) a measure of the effort required to exert the force; B) equal to F times the distance traveled; C) equal to $F \cos \theta$ times the distance traveled.

3 When the net work done on a body is negative, the body's kinetic energy: A) increases; B) decreases; C) is constant.

4 The gravitational potential energy of an object that moves horizontally: A) increases; B) decreases; C) remains constant.

5 The sum of an object's kinetic energy and gravitational potential energy will not be constant: A) if work is done by the gravitational force acting on it; B) if work is done by some force other than the gravitational force acting on it; C) unless all forces are directed vertically.

Reading 10: Energy, part II and kinetic theory (sects. 7-4 to 7-7 and 12-3)
Reading Test 10
1 Mechanical energy is conserved: A) never; B) always; C) only when nonconservative forces do no work; D) only when conservative forces do no work.

2 Power is: A) measured in kilowatt-hours; B) measured in joules; C) the change in energy; D) the rate at which work is done.

3 A spring can have potential energy: A) only when it is stretched; B) only when it is compressed; C) when it is either stretched or compressed.

4 The temperature of a container of oxygen is increased from 300K to 400K. Compared to the initial number of oxygen molecules with speed of about 900 m/s, the final number of molecules with about that speed is roughly: A) the same; B) double; C) half; D) triple.

5 According to the kinetic theory of gases, the average force exerted on the wall of a container by gas molecules is: A) proportional to the absolute temperature of the gas; B) proportional to the Celsius temperature of the gas; C) independent of the temperature of the gas.

Reading 11: Torque and static equilibrium (Sect. 9-2 & Ch. 10)
Reading Test 11
1 The moment arm of a force about an axis through a point O is the distance from O to: A) The point where the force is applied; B) any point on the line along which the force acts; C) the closest point on the line along which the force acts.

2 If a body is initially at rest and the resultant force on that body is zero: A) its center of mass won't move, but the body might rotate; B) the body won't rotate, but its center of mass might move; C) the body is in static equilibrium – its center of mass won't move and the body won't rotate.

3 In applying the condition for static equilibrium, when you calculate torque: A) you must choose an axis through the center of mass; B) you must choose an axis such that the line of action of at least one force passes through the axis; C) any choice of axis is valid; choose whatever is convenient.

4 A body's center of gravity is: A) a point where, for the purpose of computing torque, you can think of the body's entire weight acting; B) a place (such as the center of the earth) where you can place a body so that the force of gravity cancels out.

5 When you hold a heavy weight in your hand with your arm outstretched, the force exerted by muscles on bones within your body: A) is always equal in magnitude to the weight; B) is always less than the weight; C) may be much greater than the weight.

Reading 12: Fluids (Ch. 11, sects. 1 – 3 & 4 – 7)
Reading Test 12
1 In a glass jar full of water, the force exerted by the water on a very small section of glass on the side of the jar very close to the bottom, compared with the force exerted by the water on a section of glass of the same size on the bottom of the jar, is: A) smaller; B) greater; C) the same.

2 The buoyant force on a body immersed in a liquid has the same magnitude as:
A) the weight of the body; B) the weight of the liquid displaced by the body;
C) the difference between the weights of the body and the displaced liquid;
D) the average pressure of the fluid times the surface area of the body.

3 According to the continuity equation, when the cross-section of a flow tube decreases, the speed of the fluid: A) increases; B) decreases; C) stays the same; D) goes to zero.

4 Blood pressure in a major artery in the neck is: A) greatest if you are standing upright; B) greatest if your body is horizontal; C) greatest if you are upside down, doing a handstand; D) the same in all cases.

5 The overall drop in blood pressure from arteries to veins is mainly due to:
A) turbulence; B) capillarity; C) fluid friction.

Second Semester

Reading 1: Coulomb's law and the electric field (sects. 17-1 to 17-3)
Reading Test 1
1 If point P is midway between two charges of equal magnitude, a positive charge to the right of P and a negative charge to the left of P, what is the direction of the electric field (if any) at point P? A) right; B) left; C) no field; the fields of the two charges cancel.

2 Is an electric field present at a point in space even if there is no test charge at that point to measure the field? A) yes; B) no.

3 The force exerted on charge 1 by charge 2 is proportional to: A) the sum of charge 1 and charge 2; B) the difference of charge 1 and charge 2; C) the product of the magnitudes of charge 1 and charge 2.

4 The force exerted on charge 1 by charge 2 is proportional to: A) the distance between 1 and 2; B) the square of the distance between 1 and 2; C) the inverse of the distance between 1 and 2; D) the inverse of the square of the distance between 1 and 2.

5 If you know the force $\mathbf{F}_{13}$ that charge 1 exerts on charge 3 and the force $\mathbf{F}_{23}$ that charge 2 exerts on charge 3, you can find the resultant force on charge 3 by:
A) adding the magnitudes of $\mathbf{F}_{13}$ and $\mathbf{F}_{23}$; B) finding the vector sum of vectors $\mathbf{F}_{13}$ and $\mathbf{F}_{23}$.

Reading 2: Continuous charge distributions (sects. 17-4 & 17-5)
Reading Test 2
1 The direction of an electric field is: A) tangent to a field line; B) perpendicular to a field line; C) in general, neither tangent nor perpendicular to a field line.

2 Electric field lines: A) never cross; B) always cross; C) sometimes cross.

3 Electric field lines: A) begin or end only at points where there is electric charge; B) never begin or end. They are continuous everywhere.

4 For a statically charged conductor, the electric field is: A) always zero on the surface of the conductor; B) always zero inside the conductor; C) always zero outside the conductor.

5 Field lines indicate an electric field's: A) direction only; B) magnitude only; C) magnitude and direction.

Reading 3: Electric potential (sect. 18-1)
Reading Test 3
1 Electric potential is not: A) the work done by a test charge; B) a function of position in space; C) the potential energy of a test charge divided by the test charge; D) produced by a distribution of charge.

2 The source of electric potential is: A) the test charge used to measure potential; B) a distribution of electric charge; C) mass; D) none of the above.

3 An initially stationary positive charge in an external electric field will move in the direction of: A) lower potential; B) higher potential.

4 An equipotential surface is always: A) a surface along which the change in potential per unit length is the same in all directions; B) a surface along which the potential is constant; C) a spherical surface; D) a conducting surface.

5 If each of a set of point charges q_i contribute a potential V_i to the total potential V at a certain point P, the value of V is found by adding: A) vectors: sum vectors $\mathbf{V}_i$; B) magnitudes: sum magnitudes of the V_i; C) values: sum values of the V_i.

Reading 4: Capacitance (sects. 18-2 & 18-3)
Reading Test 4
1 The charges on the two plates of a capacitor are: A) always equal in sign, but may have different magnitudes; B) always opposite in sign, but may have different magnitudes; C) always equal in magnitude and opposite in sign; D) always equal in magnitude and in sign.

Reading Tests

2 One charges a capacitor by: A) rubbing one of the plates against the right material, utilizing static electricity; B) using wires to connect the capacitor to the terminals of a battery or power supply.

3 Placing a dielectric between the plates of a charged, isolated capacitor causes the magnitudes of the charges on the plates of the capacitor to: A) decrease; B) increase; C) remain unchanged.

4 Placing a dielectric between the plates of a charged, isolated capacitor causes the electric field between the plates of the capacitor to: A) increase; B) remain unchanged; C) decrease.

5 Placing a dielectric between the plates of a charged, isolated capacitor causes the potential difference between the plates of the capacitor to: A) decrease; B) remain the same; C) increase.

Reading 5: Current; Ohm's law (Ch. 19)
Reading Test 5
1 Electric current in a wire that is part of a curcuit is produced by the motion of: A) neutrons; B) protons; C) electrons.

2 To light a light bulb: A) electric current must flow through the bulb's filament; B) electric current must flow through a wire touching the base of the bulb; C) none of the above.

3 A battery is being used to operate a flashlight. Inside the battery: A) chemical energy is converted to electrical energy; B) heat is converted to electrical potential energy; C) chemical energy is converted to kinetic energy; D) electrical energy is converted to chemical energy.

4 Ohm's law relates: A) capacitance, charge, and voltage; B) the charge on a conductor to the electric field it produces; C) the current through a body to the potential difference across the body.

5 Inside a resistor, electrical energy: A) is converted to chemical energy; B) is converted to thermal energy; C) remains unchanged.

Reading 6: DC circuits (sects. 20-1, 20-2, 20-3, 20-5, & 20-7)
Reading Test 6
1 When the filament of a light bulb breaks and the light goes out, this is an example of: A) an open circuit; B) a short circuit; C) a steady current; D) a closed circuit.

2 When you go around a circuit from a point A and back to the same point A, the sum of the voltage drops is: A) dependent on the path; B) zero; C) negative; D) positive.

3 Resistors are in parallel if they are: A) connected so that they necessarily have the same voltage across each resistor; B) connected so that they necessarily have the same current in each resistor; C) oriented along parallel lines.

4 Resistors are in series if they are: A) placed in a straight line; B) connected so that the voltage across each resistor is necessarily the same; C) connected so that the current through each resistor is necessarily the same.

5 The equivalent resistance of a network of resistors is always: A) the resistance of a single resistor that carries the same current as the network when the same voltage is applied to it; B) the inverse of the sum of the inverse resistances of the resistors in the network; C) the sum of the resistances of the resistors in the network.

Reading 7: Magnetism (sects. 21-1 to 21-4)
Reading Test 7
1 The source of a magnetic field is: A) moving charge; B) stationary charge; C) none of the above.

2 The effect of a magnetic field is to exert force on: A) moving charge; B) stationary charge; C) none of the above.

3 A charge in a magnetic field experiences a force: A) if it is moving in any direction; B) if it is moving with a component of velocity perpendicular to the field; C) always.

4 The magnetic force on a current carrying wire in a magnetic field is directed: A) perpendicular to the wire and perpendicular to the field; B) parallel to the wire in the direction of negative current; C) perpendicular to the wire and parallel to the field; D) parallel to the wire in the direction of positive current.

5 Magnetic field lines produced by current in a long straight wire are: A) circles centered on the wire; B) directed radially out from the wire; C) directed radially in toward the wire.

Reading 8: Mechanical waves and light (sects. 16-1 & Ch. 23)
Reading Test 8
1 Within a laser beam there must be: A) both an electric field and a magnetic field; B) a magnetic field, but not an electric field; C) neither an electric field nor a magnetic field; D) an electric field, but not a magnetic field.

2 Which, if any, of the following do NOT travel through a vacuum at a speed of 3.0×10^8 m/s? A) gamma rays; B) microwaves; C) radar; D) X-rays; E) none of the above; all travel at that speed.

3 What kind of waves require a material medium through which to move? A) microwaves; B) radio waves; C) light waves; D) sound waves.

4 What determines the frequency of a wave? A) the wavelength; B) the wave source; C) the wave speed; D) the medium through which it moves.

5 The index of refraction of a material is related to: A) the angle at which light reflects off of it; B) how fast light travels through it; C) how hard the material is; D) the angle at which light is incident on it.

Reading 9: Plane mirrors and thin lenses (sects. 24-1 & 24-3)
Reading Test 9
1 A high power lens has: A) a long focal length; B) a short focal length.

2 An object's image in a plane mirror is located: A) in front of the mirror; B) on the mirror's surface; C) either in front of or behind the mirror, depending on how far the object is from the mirror; D) behind the mirror.

3 When incident parallel light rays in air pass from air into glass with a convex front surface, inside the glass the light rays: A) neither converge nor diverge; B) converge; C) diverge.

4 When incident parallel light rays in air pass from air into glass with a concave front surface, inside the glass the light rays: A) diverge; B) neither converge nor diverge; C) converge.

5 An object's image formed by a positive thin lens will be bigger if the image is: A) closer to the lens than the object; B) farther from the lens than the object; C) neither closer nor farther; distances don't affect image size.

Reading 10: The human eye (sects. 25-1 to 25-3)
Reading Test 10
1 When you see an object: A) a real image is formed on your cornea; B) a virtual image is formed on your retina; C) a real image is formed on your retina; D) a virtual image is formed on your cornea.

2 The eye's crystalline lens: A) is spherical; B) provides most of the eye's optical power; C) is flexible, variable in shape; D) is rigid, fixed in shape, but not spherical.

3 We can see objects clearly over a range of object distances because our eyes adjust: A) the refractive index of the medium inside of the eye; B) the curvature of a surface inside the eye; C) the curvature of the cornea; D) the length of the eyeball.

4 When an optical instrument produces an angular magnification > 1: A) this means the same as a linear magnification >1; B) this is not related to the size of the image on the retina; C) the size of the image on the retina is increased; D) the size of the image on the retina is decreased.

5 You can use a magnifier to improve your image of: A) an atom; B) an ant; C) an object that is far away; D) an object 10^{-4} mm in diameter.

Reading 11: Interference (sects. 26-1 & 26-2)
Reading Test 11
1 The bending of light around an obstacle is called: A) interference; B) diffraction; C) coherence; D) resolution.

2 Light in a laser beam can be polarized: A) in any direction whatsoever; B) in any direction parallel to the beam; C) in any direction perpendicular to the beam.

3 Interference: A) is always constructive; B) is always destructive; C) may be either constructive or destructive; D) none of the above.

4 At a certain point in a laser beam, light observed at two different times will: A) always be coherent; B) never be coherent; C) be coherent only if the time interval between the observations is not too great.

5 Light from a laser passes through a thin slit of variable width and hits a screen a few meters away. As the slit gets narrower, the width of the light on the screen: A) decreases; B) does not change; C) increases.

Reading 12: Diffraction and polarization (sects. 26-3 & 26-4)
Reading Test 12
1 Initially unpolarized light is incident on a polarizer. Compared to the incident intensity of light, the intensity of light transmitted is: A) 50%; B) 25%; C) 75%; D) 100%.

2 Initially unpolarized light can be polarized: A) only by scattering; B) only by absorption; C) only by reflection; D) by reflection, absorption, or scattering.

3 A diffraction grating is: A) a special type of microscope; B) a covering used to protect a lens; C) a glass plate with thousands of closely spaced slits.

Reading Tests

4 When two point sources of light images are overlapping Airy disks with the center of one's Airy disk at the edge of the other's Airy disk, according to the Rayleigh criterion: A) the points are unresolved; B) the points are clearly resolved; C) the points are barely resolved.

5 The image produced on distant screen by a plane light wave passing through a very narrow slit is: A) a narrow band of light; B) a central broad band of light with narrower, dimmer bands of light on either side; C) a central narrow band of light with broader, dimmer bands of light on either side; D) a series of equally wide, equally bright narrow bands of light.

Reading 13: Relativity (sects. 27-1 & 27-2)
Reading Test 13
1 Science fiction stories in which a space traveler does not age as much as those who remain on earth: A) do have a scientific basis; B) do not have a scientific basis.

2 The laws of physics are valid in: A) any inertial reference frame; B) any reference frame; C) any noninertial reference frame.

3 Has Einstein's time dilation formula been confirmed experimentally?
A) yes; B) no.

4 The speed of light in vacuum: A) is independent of both the motion of the observer and the motion of the light source; B) depends on the motion of the light source; C) depends on the motion of the observer; D) depends on where you measure it.

5 The speed of sound: A) is independent of the reference frame with respect to which the speed is measured; B) depends on the reference frame with respect to which the speed is measured.

Reading 14: Quantum mechanics (Ch. 28 & Ch. 29)
Reading Test 14
1 For which color light do photons have the least energy? A) red; B) blue; C) green; D) yellow.

2 Light shines on a metal in a photocell and electrons are emitted. The maximum kinetic energy of the electrons depends on: A) only the intensity of the light; B) only the frequency of the light; C) both the intensity and the frequency of the light; D) neither the intensity nor the frequency of the light.

3 When an X-ray scatters off an electron, imparting energy to the electron, the X-ray's frequency: A) increases; B) remains the same; C) decreases.

4 The colors of an atom's spectral lines are an indication of the atom's: A) allowed energy levels; B) temperature; C) momentum.

5 The fact that all the electrons in a multi-electron atom are not in the very lowest electron energy state is a manifestation of the fact that: A) the electrons won't all fit in a very small space; B) the atom is not at a temperature of absolute zero; C) no two electrons can have all the same quantum numbers.

10

Clicker Questions

"Begin challenging your own assumptions. Your assumptions are your windows on the world. Scrub them off every once in awhile, or the light won't come in."

Alan Alda

First Semester

Week One: Introduction to Thinking in Physics

1 Do you have a strong interest in learning physics? A) yes; B) no.

2 Do you have anxiety about taking this course? A) yes; B) no.

3 Is a force required to keep an object moving? A) yes; B) no.

4 Everyone has some concepts related to force and motion that work well in everyday life, but that have limitations. How can we change these preconceptions? Will your concepts change as a result of having the correct concepts explained to you? A) always; B) usually; C) usually not.

5 Can you learn physics by only paying close attention in class while everything is carefully explained to you? A) yes; B) no.

6 If your overall average at the end of this course were 80%, what would be your grade in this class? A B C

7 When you first look at a problem on a physics test, should you be able to know how to solve it? A) yes; B) no.

Concepts Class 1: Measurement, Motion, Vectors

1 OK, we're about to jump off this bridge with a bungee cord attached to our feet. The ground is 85 feet below. The bungee cord is 30 yards long, which is 60 feet. Great! Let's jump. This analysis will result in: A) a safe jump; B) certain death.

2 The height of someone in this class is: A) 66 in; B) 66.0012 in.

3 The length of this room, measured as carefully as you can, is: A) 10 m; B) 10.1 m; C) 10.1437 m.

4 When you multiply a number with two significant figures by a number with four significant figures, how many significant figures does the answer have?
A) 1; B) 2; C) 3; D) 4.

5 The sum of 3.2 and 1.43 is: A) 4; B) 4.6; C) 4.63.

6 You are calculating the total floor area of two rooms, with dimensions 4.15 m by 4.63 m and 3.8 m by 3.25 m. Your calculator reads 31.5645. How should you express this area? A) 31.5645; B) 31.6 m^2 ; C) 32 m^2; D) 31.6; E) 31.6 m.

7 Where is the speed of the club head greatest?

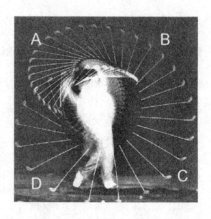

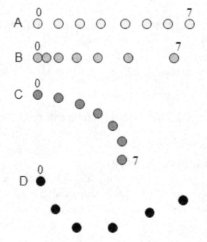

Disks A,B,C, & D are shown at 1 s intervals from $t = 0$ to $t = 7$ s.

8 Which disk has the greatest average speed over the 7 s?

9 Which disk has the greatest instantaneous speed, and during what time interval does it occur?

Each dart in the figure below represents a displacement vector of magnitude 10 m.

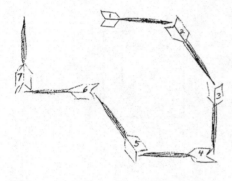

10 Which, if any, of the vectors are the same? A) none; B) 1 and 4; C) 2 and 5; D) 4 and 6; E) 3 and 7.
11 What is the resultant? A) 10 m; B) 10 m to the right; C) 10 m to the left; D) 10 m down; E) 70 m.
12 Suppose you reverse the order in which you connect the darts: start with 7, then place 6 with its tail at the tip of 7, and so on. The resultant vector would then be: A) the opposite, 10 m to the right; B) the same, 10 m to the left; C) no simple answer; you have to construct.
13 Suppose you rotated each of these vectors 180 degrees. The vector sum would then be: A) the same as before; B) the same in magnitude, but opposite in direction; C) different in both magnitude and direction.

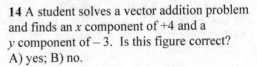

14 A student solves a vector addition problem and finds an *x* component of +4 and a *y* component of − 3. Is this figure correct? A) yes; B) no.

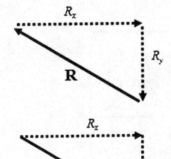

15 A student solves a vector addition problem and finds an *x* component of +4 and a *y* component of − 3. Is this figure correct? A) yes; B) no.

16 A student solves a vector addition problem and finds an *x* component of +4 and a *y* component of − 3. Is this figure correct? A) yes; B) no.

Concepts Class 2: Acceleration, Relative Motion
1 How many devices in a car are used to produce acceleration? A) one; B) two; C) three or more.

2 You are walking at constant speed along the corridors of an office building. You turn 90 degrees to the right. Which of the sketches below shows the correct vector representing your change in velocity, if your initial velocity is in the positive *y* direction?

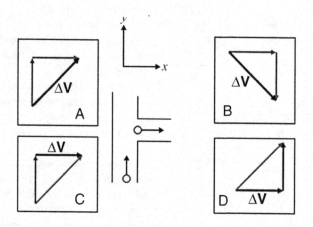

3 You are walking across campus at constant speed, and you change direction, turning 45 degrees to the left. Which of the sketches below shows the correct vector representing your change in velocity, if your initial velocity is in the positive *y* direction?

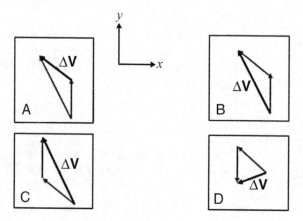

4 This car is moving north, relative to earth, at 30 mph. What is the velocity of the earth, relative to the car?
A) The earth does not move; B) 30 mph, north; C) 60 mph, north; D) 30 mph, south; E) 45 mph, east.

Video Concept Question

5 Joe is standing on the sidewalk on Loyola Blvd. as you pass by in your truck, traveling south at 100 km/h. You shoot a ball from the back of your truck, giving it an initial velocity of 100 km/h, directed horizontally, north. Describe the motion of the ball, as Joe sees it.

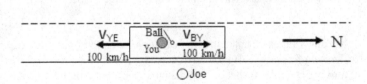

A) travels in a straight line at constant velocity of 100 km/h, north;
B) travels in a straight line at constant velocity of 100 km/h, south;
C) travels as a projectile with an initial horizontal velocity of 100 km/h, north;
D) travels as a projectile with an initial horizontal velocity of 100 km/h, south;
E) free falls from rest.

See the video of this experiment (performed in Japan) on YouTube.
http://www.youtube.com/watch?v=yPHoUbCNPX8

Concepts Class 3: Newton's Laws of Motion

1 The Great Meteor Crater in Arizona was formed about 50,000 years ago when a very large meteoroid struck the earth. If such a meteoroid were to hit the earth in the future, could the location of impact be predicted if the meteoroid were observed months before it hit?
A) yes; B) no.

Video Concept Question
2 Hundreds of shopping carts are wheeled into the back of a large truck at a loading dock. The truck's back door is not closed. What happens to the carts when the truck slowly moves away from the dock? A) The carts stay in place inside the truck and move with the truck; B) The carts do not move with the truck. See the answer in this YouTube video: http://www.youtube.com/watch?v=x_4TqtHATWI

3 Is Newton's first law satisfied in all reference frames? A) yes; B) no.

4 If you ride a bicycle in a straight line at constant speed, are you in an inertial reference frame? A) yes; B) no.

5 If you ride a merry-go-round, are you in an inertial reference frame? A) yes; B) no.

6 In which of the following cases can a moving car serve as an inertial reference frame? i) accelerating from rest; ii) moving at constant speed along a straight, level road; iii) moving at constant speed around a curve; iv) rounding the top of a hill at constant speed; v) slowing down, as the brakes are applied. A) yes; B) no.

7 You apply the brakes on your car, stopping suddenly, and are thrown forward. What force, if any, acting directly on you, is responsible for your forward motion? A) the force of the brakes; B) the force of the road; C) the force of the seat back; D) the force of the acceleration; E) no force.

8 A novice ice skater who has not yet learned how to stop is headed straight into the rail at the edge of an ice rink. What kind of force, if any, keeps the skater moving? A) the force of friction; B) the force of the ice; C) the force of the skater; D) the force of momentum; E) no force.

9 Which has more mass, a bowling ball or a billiard ball?
A) bowling ball; B) billiard ball.
How could you demonstrate that your answer is correct?

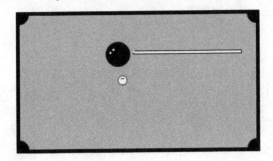

10 Is the particle shown in this figure accelerated? If so, in what direction?
A) yes, upward; B) yes, downward; C) yes, but not straight up or straight down; D) no.

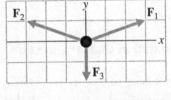

11 Is the particle shown in this figure accelerated? If so, in what direction?
A) yes, upward; B) yes, downward; C) yes, to the right; D) yes, to the left; E) no.

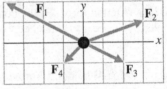

12 After pushing his arm backward through the water, a freestyle swimmer lifts his arm out of the water to bring it forward. Why does he move his arm forward through air rather than through water? A) to get his hand forward faster; B) to get an extra push as his hand falls down into the water; C) to prevent pushing himself backward.

13 A bug hits the windshield of a car you are driving. Is the force exerted on the bug greater than, equal to, or less than the force exerted by the bug on the car? A) greater than; B) equal to; C) less than.

A spacecraft has engines on 4 sides which can be turned on or off. Initially the spacecraft is moving to the right at constant velocity. Turning on the engine on the left side exerts a force to the right on the craft, accelerating it to the right.

14 If both left and right engines are on, as shown here, what motion results? A) slows down; B) continues at constant velocity; C) changes direction.

15 If only the bottom engine is on, as shown here, what motion results? A) immediately begins moving in positive y direction; B) curves increasingly upward toward the positive y direction.

Demonstration Clicker Question

16 When I push on the blackboard, why doesn't it move? A) the force I exert is cancelled by a reaction force; B) the board is too massive; C) the board is attached to the building.

Followup question: Why don't I move?

Concepts Class 4: Force Laws, Applications of Newton's Laws

1 Calculate your weight in N and your mass in kg. (1 lb = 4.45 N) Compared to the number you just computed, your weight while flying in a commercial aircraft is: A) greater than; B) less than; C) the same. Assume your mass is unchanged.

Clicker Questions

2 What forces are acting on this man who weighs 200 lbs? A) no forces; B) only his weight, a downward force of 200 lbs; C) an upward normal force of 200 lbs; D) his weight and a normal force, an action-reaction pair; E) his weight (the pull of the earth) and a normal force (the push of the road).

3 In which of these cases, if any, do people have a weight that differs significantly from their normal weight (either less than or greater than normal weight)? A) cheerleaders standing on each other's shoulders; B) a person falling into a swimming pool after jumping off a high dive; C) none;

D) floating in a swimming pool;

E) astonauts in training floating in an airplane while it is falling.

4 The cables supporting the Golden Gate Bridge are under enormous tension. At point P, the section to the right of P exerts a force **T** on the section to the left of P. What is the reaction force to this force? A) a force that the section of cable to the left of P exerts on the section of cable to the right of P; B) a force that holds the cable to the left of P in place.

The cable consists of 27,572 strands of flexible wire.

5 A heavy blanket hangs from a clothesline. Will the tension in the clothesline be greater if it sags a little or if it sags a lot?
A) sags a little; B) sags a lot;
C) same in either case.

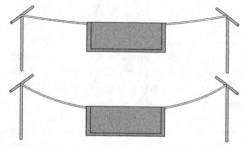

6 The hammock is supported by two ropes. A) Tension T_1 is greater than tension T_2; B) Tension T_2 is greater than tension T_1; C) The tension is the same in both ropes

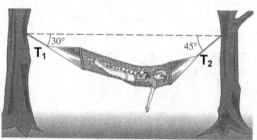

Concepts Class 5: Projectiles

Questions 1—5 A projectile of mass m near the surface of the earth is given an initial velocity $\mathbf{v}_0$. Nothing is touching the mass except air, and air resistance is negligible.

1 Can the motion of the projectile ever depend on its mass?
A) yes; B) no; C) sometimes.

2 Does v_x ever change during the projectile's motion?
A) yes; B) no; C) sometimes.

3 Does v_y ever change during the projectile's motion?
A) yes; B) no; C) sometimes.

4 Can v_y ever increase during the projectile's motion?
A) yes; B) no; C) sometimes.

5 Can the change in v_y during one second of its motion ever vary?
A) yes; B) no; C) sometimes.

6 A rock is thrown from a hill with an initial velocity of 10 m/s, directed horizontally. About how long does it take for the rock's velocity to be directed 45 degrees below the horizontal? A) 0.5 s; B) 1 s; C) 1.4 s; D) 2 s; E) impossible to determine without more information.

Clicker Questions

7 A football is kicked with horizontal and vertical components of initial velocity both equal to 20 m/s. **a)** About how long is the ball in the air? A) 1 s; B) 2 s; C) 4 s; D) 8 s; E) 16 s. **b)** What is its approximate range? A) 5 m; B) 10 m; C) 20 m; D) 40 m; E) 80 m.

8 You toss a calculator to a classmate across the room. At the instant the calculator is at the highest point, its acceleration is: A) 0; B) 9.8 meters per second squared, directed down; C) 9.8 meters per second squared, directed horizontally.

Concepts Class 6: Friction

Demonstration Clicker Questions: 1 – 7

1 A book is initially at rest on a table. We know that its weight must be balanced by a force **S** exerted on the book by the surface of the table. Now suppose you push the book slightly, exerting a small horizontal force **P** on the book to the right, but not big enough to make the book move. Consider whether the other forces on the book change? **a)** Does the weight change? A) yes; B) no. **b)** Does the force **S** exerted by the surface change? A) yes; B) no.

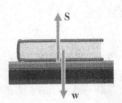

2 This book weighs 10 N and to move it along the surface of the table, a minimum horizontal force of 3 N must be applied. If I push down on the top of the book with my hand with a force of 100 N, the minimum horizontal force required to move the book is: A) 3 N; B) 10 N; C) 30 N; D) 100 N; E) 33 N.

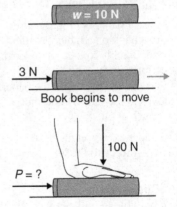

I hold a book at rest against a wall, exerting a horizontal force of 20 N. The book weighs 10 N.

3 The normal force exerted by the wall on the book is directed: A) up; B) down; C) in – toward the wall; D) out – away from the wall.

4 This normal force has magnitude: A) 10 N; B) 20 N; C) 30N.

5 The friction force exerted by the wall on the book is directed: A) up; B) down; C) in; D) out.

6 This friction force has magnitude: A) 10 N; B) 20 N; C) 30 N.

7 The static coefficient of friction between wall and book is: A) 0.5; B) 1.0; C) 2.0; D) undetermined.

8 The sidewalk exerts on the foot of this walking person: A) a friction force to the right; B) a friction force to the left; C) no friction force.

Demonstration Clicker Questions: 9, 10

9 A sheet of paper is initially at rest beneath a book on a table. You jerk the paper quickly to the right, exerting a large force on it. The book:
A) moves to the right with the paper;
B) moves to the left; C) does not move much at all.

10 The paper exerts on the book: A) a friction force to the right; B) a friction force to the left; C) no friction force.

Video Clicker Question

11 Is it possible to pull a tablecloth off of a table, without moving the dishes that were on top of the tablecloth? A) yes; B) no.

YouTube instructional video showing how to do it:
http://www.youtube.com/watch?v=jt0pco9taQg&feature=fvsr

Clicker Questions

12 This truck is moving at constant speed along a straight, level road. The crate does not slide. Is there a friction force exerted by the truck bed on the crate?
A) yes, directed to the right; B) yes, directed to the left; C) no.

13 Which of the following is always correct for the magnitude of the normal force exerted by a surface on an object in contact with the surface? A) always equals the weight of the object; B) always equals the component of the weight perpendicular to the surface, $mg \cos \theta$; C) always is whatever it needs to be to keep the object from going through the surface.

14 Two identical skiers on identical skis are skiing down slopes that are identical, except for steepness. Slope B is much steeper than slope A. The friction force exerted by the snow on the skis is: A) greater for A; B) greater for B; C) zero for both A & B; D) the same for A & B, but not zero.

Concepts Class 7: Circular Motion, Center of Mass

1 A car rounds a curve in the road. What body exerts a force on the car, causing its acceleration?
A) the car's steering wheel; B) the surface of the road; C) the entire earth.

2 A car goes over the crest of a hill. What body exerts a force on the car, causing its acceleration? A) the car's steering wheel; B) the surface of the road; C) the entire earth.

3 Inside a washing machine is a cylindrical drum, perforated with holes. During the spin part of the wash, the drum rotates rapidly and much of the water in the clothes passes out through the holes in the drum. What force, if any, causes the water to leave the clothes? A) an outward force; B) an inward force; C) normal force; D) weight; E) no force.

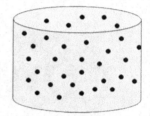

4 Which of the paths shown below does a water drop follow as it exits the drum, which is spinning counterclockwise?

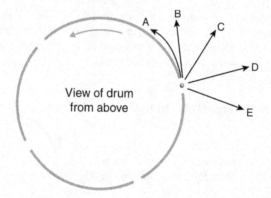

5 Why don't the clothes leave the drum? A) the clothes weigh more than water drops; B) the clothes have more inertia than the water drops; C) the holes are small, clothes are large, and the normal force of the drum holds the clothes in; D) the drum does not rotate fast enough to cause the clothes to leave the drum.

A Ferris wheel is rotating counterclockwise in a vertical circle.

6 The resultant force on a rider on a Ferris wheel at point P is: A) in the same direction as the rider's weight; B) in the opposite direction of the weight; C) some other direction.

7 The force exerted by the seat on the rider is: A) greater than the rider's weight; B) less than the rider's weight; C) equal to the rider's weight.

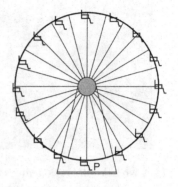

8 S is the magnitude of the force exerted by the seat on the rider, and w is the magnitude of the rider's weight. The resultant force on the rider at point P is: A) $S + w$; B) $S - w$; C) $w - S$; D) $-S - w$; E) S.

Demonstration Clicker Questions

9 Anatol is going to place a glass of water on the platform, which is attached to strings, and then attempt to swing the platform in a vertical circle with the glass staying in place on the platform and without spilling any water out of the glass. Will he succeed? A) yes; B) no.
Demonstrate
Explain why, when the water glass is upside down at the top of the circle, neither the glass nor the water falls down.

10 Consider the forces acting on the glass at the instant it was at the position shown in this figure, moving in the counterclockwise direction. Assume that the force of friction on the glass is negligible. What are the forces on the glass and their effects? A) weight and normal force, which balance, giving zero net force; B) only weight, which provides centripetal acceleration; C) only normal force, which provides centripetal acceleration; D) normal force, which causes a change in direction of motion, and weight, which causes the glass to slow down; E) normal force and weight, which both contribute to centripetal acceleration.

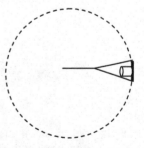

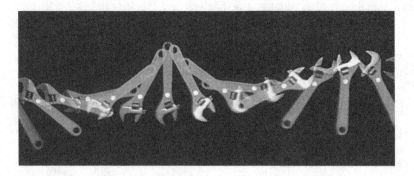

11 A wrench rotates as it slides along a horizontal surface. Its center of mass is marked by a white spot. What can you conclude about forces acting on the wrench? A) a horizontal force acts on the wrench, causing it to spin; B) there are no forces acting on the wrench; C) there are no forces on the wrench other than its weight; D) friction is the only force on the wrench; E) the resultant force on the wrench is zero. Explain.

12 Two identical bricks are stacked as shown. The center of mass of the system is closest to which point?

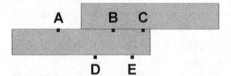

13 Suppose you are initially at rest on a perfectly frictionless ice pond. Would it be possible for you to walk or crawl very carefully to shore? A) yes; B) no.

14 A spacecraft is initially far from earth in interstellar space, and is at rest with respect to earth. Is it possible for the center of mass of the spacecraft and its entire contents to move, if no external body exerts a force on the system? A) yes; B) no.

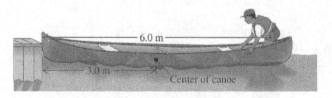

15 Suppose you are in the back of a canoe and the front end has just touched shore. You want to move to the front of the canoe and step out onto shore. What happens when you move forward in the canoe? A) the canoe pushes forward into shore; B) the canoe moves away from shore; C) the canoe does not move.

Concepts Class 8: Gravitation, Momentum

1 Newton wondered: the apple falls from the tree; why doesn't the moon fall? A) the moon does accelerate towards earth, you just don't notice it; B) the moon doesn't fall because the resultant force on the moon is zero; C) terrestrial gravity doesn't act on objects in space; D) the force of the sun on the moon prevents it from falling.

Moon

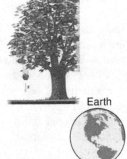
Earth

2 Earth exerts a gravitational force of 2×10^{20} N on the moon. The force exerted by the moon on the earth is: A) less; B) greater; C) the same.

3 Suppose the masses of two particles were both increased to three times their original values. By what factor would the gravitational force between them change? A) 3; B) 6; C) 8; D) 9; E) 27.

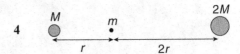

Questions 4 – 6
The direction of the resultant force on the body of mass *m* in each figure is: A) right; B) left; C) up; D) down;

E)

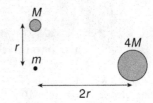

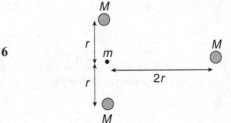

7 Suppose a planet has a radius twice as large as the earth's radius but the same mass. How much would an object weighing 100 N on earth weigh on this planet? A) 100 N; B) 200 N; C) 50 N; D) 25 N.

8 Suppose you weigh 800 N at the surface of the earth. What would your weight be if you were at the center of the earth? A) 0; B) 800 N; C) 400 N; D) 200 N; E) infinite.

9 Suppose you are an astronaut, standing on the surface of the moon and taking target practice with a rifle aimed horizontally at a target. The rifle you are using is a newly developed kind that fires bullets at a very high velocity – 1.7 km/s (roughly 60% faster than conventional high velocity bullets like the .240 Weatherby). A fellow astronaut warns you that you are in danger of missing the target and hitting yourself in the back of the head with a bullet that orbits the moon, just above the surface. Is your fellow astronaut correct?   A) yes; B) no.
The orbital velocity of a lunar satellite is 1.7 km/s.

Some "impulsive" forces – large forces of short duration

10 After being hit, each of the balls in the photos above has momentum. Values range from 3 kg-m/s to 10 kg-m/s. In addition to the momentum of each ball, what else do you need to know to estimate the average force that acted on the ball?
A) compression of ball; B) direction of hit; C) time of contact.

11 A hockey puck has an initial momentum directed north. Then the puck is hit with a stick, exerting on the puck a force directed east and providing an impulse to the puck equal in magnitude to the puck's initial momentum. What is the direction of the puck's momentum after the hit? A) N; B) E; C) S; D) NE; E) SE.

Clicker Questions

12 On January 1st the earth has momentum **p** relative to the sun.
a) By how much does the earth's momentum change in one year?
A) **p**; B) 2**p**; C) – **p**; D) – 2**p**; E) 0.
b) By how much does the earth's momentum change in six months?
A) **p**; B) 2**p**; C) – **p**; D) – 2**p**; E) 0.

13 No significant *external* force acts on a shotgun as it is being fired. What can one conclude about the total momentum of the system of the shotgun and the shot after the gun is fired? A) depends on the mass of the shotgun; B) depends on the velocity of the shot; C) depends on the mass of the shot; D) always is equal to zero.

14 Ideally, is it a good idea for a shotgun to be as light as possible?
A) yes; B) no. Explain.

15 Suppose you are at a shooting gallery at an amusement park, firing pellets from an air rifle at targets that fall over if hit hard enough in the right place. Assuming all pellets have the same mass and initial velocity, which should be more effective in knocking over targets? A) pellets that are absorbed by the target; B) pellets that pass through the target; C) pellets that hit the target and drop down; D) rubber pellets that bounce off the target.

Concepts Class 9: Energy, part one
1 A weightlifter holds a heavy barbell at rest. Is any work done by either of the forces, **F** or **w**, exerted on the barbell?
A) no; B) yes, work is done by both of the forces; C) yes, work is done by **F**, but not by **w**; D) yes, work is done by **w**, but not by **F**.

2 A skater moves, as vertical forces, **w** and **N**, act on her. Is any work done by either of the forces on the skater?
A) no; B) yes, work is done by both of the forces; C) yes, work is done by **N**, but not by **w**; D) yes, work is done by **w**, but not by **N**.

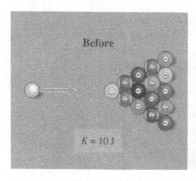 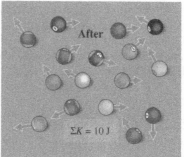

3 In a game of pool, the kinetic energy of the cue ball is initially 10 J, and all of the other balls are initially at rest. After the break, the sum of the kinetic energies of all the balls is 10 J. Is kinetic energy conserved for this system? A) yes; B) no.

4 A child swings on a rope hanging from a tree. Which, if any, of the forces on the child do *no* work? A) weight; B) tension; C) resultant force; D) none.

A cue stick strikes a cue ball that is initially at rest, exerting a force **F** on the ball as it begins to move.

5 The work done by **F** is: A) positive; B) negative; C) zero.

6 The change in the cue ball's kinetic energy is: A) positive; B) negative; C) zero.

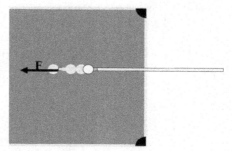

7 In bungee jumping, a person leaps off a bridge, with one end of a bungee cord attached to the feet and the other end attached to the bridge. When the bungee cord begins to stretch, while you are still moving downward, the work done by the tension force of the cord on your body is: A) positive; B) negative; C) zero. What effect does this have on your kinetic energy?

Clicker Questions

Questions 8, 9 A car has an initial kinetic energy of 200,000 J, and then it stops as the brakes are applied.

8 What is the net work done on the car as it is stopping? A) 200,000 J; B) –200,000 J; C) 0; D) 100,000 J.

9 Which of the following six forces may contribute to the net work on the car? (Don't make any assumptions about the road, and don't idealize the motion.)
1) normal force exerted by the road; 2) friction exerted by the road; 3) weight of car; 4) force of the driver's foot on the brakes; 5) force of the brake drums on the wheels; 6) force of air resistance on car.
A) combination of 1, 2, & 3; B) combination of 2, 3, & 6; C) combination of 2 & 6; D) Only 2; E) all six forces.

10 A mountain climbing expedition begins at a base camp, climbs Mt. Everest, and days later returns to base camp. How much work is done by gravity on the body of an expedition member for the complete trip? A) a large positive number; B) a large negative number; C) 0.

11 A grasshopper jumps off the ground at an initial speed of 3 m/s. At what speed is the grasshopper moving just before it lands on the ground? A) 3 m/s; B) > 3 m/s; C) < 3 m/s; D) – 3 m/s.

12 Suppose you are designing a new slide for a playground. To offer more excitement, you want to increase the maximum speed of a child at the bottom. How high should the new slide be, in order to double the speed at the bottom, if the old slide is 1.5 m high? Assume friction to be negligible. A) 2.1 m; B) 3 m; C) 4 m; D) 6 m.

Concepts Class 10: Energy, part two

The girl in this picture is approaching the high point of her swinging.

1 During the tenth of a second time interval just before the picture was taken, her kinetic energy was: A) increasing; B) decreasing; C) constant.

2 The net work on her body during this time interval was: A) positive; B) negative; C) zero. Explain.

A ball is attached to one end of a spring of negligible mass. The ball and spring form a pendulum of varying length, swinging about point O.

3 At which of the points A, B, C is gravitational potential energy greatest, or is it the same at all points (D)?

4 At which of the points A, B, C is spring potential energy greatest, or is it the same at all points (D)?

5 At which of the points A, B, C is total mechanical energy greatest, or is it the same at all points (D)?

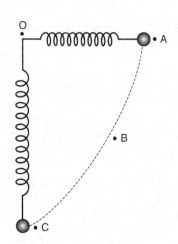

6 A book moves across a tabletop from a point i to a point f, 1 m away. Throughout the motion, a constant friction force of 2 N opposes the motion. How much work is done by the friction force? A) 0; B) 2 J; C) – 2 J; D) can't be determined unless the path from i to f is specified.

7 Can the work done by a force always be expressed as a decrease in potential energy? A) yes; B) no.

8 A boy rides a bicycle along level ground at *approximately* constant velocity, without pedaling. Is mechanical energy *approximately* conserved? A) yes; B) no.

9 A boy rides a bicycle along level ground at constant velocity, pedaling at a steady rate. **a)** Is mechanical energy conserved? A) yes; B) no.
b) Is there any work done by individual nonconservative forces? A) yes; B) no.
c) Is there any net work done by nonconservative forces? A) yes; B) no.

10 The runner and the cyclist have traveled the same distance at the same speed. Why is only the runner tired? A) the cyclist is in better condition; B) the cyclist has a machine (the bicycle) to provide energy; C) the runner wastes more energy; D) energy has nothing to do with it; the pedals provide the cyclist with leverage.

11 Hiking trails on a steep mountain usually have numerous switchbacks, back and forth winding trails that gradually go up the mountain. The reason for this is so that it takes a hiker less: A) energy; B) power; C) time to go up the mountain.

Concepts Class 11: Torque and Static Equilibrium

1 You are trying to open a door that is stuck by pulling on the doorknob in a direction perpendicular to the door. If you instead tie a rope to the doorknob and then pull with the same force, is the torque you exert increased? A) yes; B) no.

Questions 2 – 4 You are using a wrench to turn a nut, to rotate it counterclockwise around an axis through point O at the center of the nut. You apply a force of 40 N to the wrench, which may or may not be enough to turn the nut.

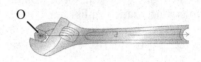

2 You should apply the force to the wrench: A) at the middle of the arm of the wrench; B) at the right end of the wrench.

3 The direction of the force should be: A) vertically upward; B) vertically downward; C) at an angle upward and to the right; D) at an angle upward and to the left.

4 Suppose the force you apply is not enough to turn the nut. Might it be possible for you to turn the nut by extending the effective length of the handle by sliding a long pipe over the handle and applying the same force as before to the right end of the pipe? A) yes; B) no. Explain.

Experiment Clicker Question

5 Equal magnitude, oppositely directed forces are applied to a body. They are the only forces acting on the body, which is initially at rest. Will the body move? A) yes; B) no; C) maybe. Try it for an object on the top of your desk.

The gate in this figure is acted upon by four forces, as shown.

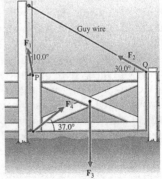

6 What is the net torque on the gate with respect to an axis through point P? A) positive; B) negative; C) zero.

7 What is the net torque on the gate with respect to an axis through point Q? A) positive; B) negative; C) zero.

8 Is this body in static equilibrium? A) yes; B) no.

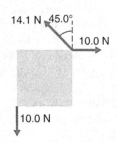

A square block is subject to the three forces shown in the figure. A fourth force is applied to the block to produce static equilibrium.
9 Find the magnitude & direction of the fourth force. A) 0 N; B) 10 N up; C) 20 N up; D) 10 N down.
10 Where is the fourth force applied to the block? A) anywhere; B) only at the center; C) anywhere along a vertical line through the center.

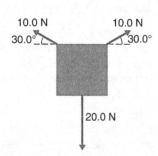

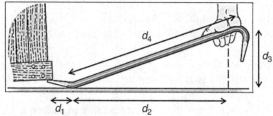

11 A heavy crate is lifted by applying a downward force of 50 N to the right end of this crowbar. To determine how much force is applied to the crate, you would need to measure which distances? A) d_1 and d_2; B) d_1 and d_3; C) d_1 and d_4; D) d_2 and d_3; E) d_2 and d_4.

Demonstration Clicker Question

12 A wine bottle is supported by a holder, a board with a hole through which the neck of the bottle is inserted. The bottom end of the holder is supported by a table. What can you conclude about the location of the center of gravity of the system of bottle and holder? A) it's directly over point P; B) it's directly over point Q; C) it's directly over point R.
13 Is it sufficient for the wine bottle holder to exert a single force on the wine bottle, applied at some point on the neck of the bottle? A) yes; B) no. Explain.

Clicker Questions

14 A long pole is held horizontally with two hands. If the pole is held with both hands near one end, the magnitude of the force exerted by each hand will be:
A) greater than the weight of the pole; B) less than the weight of the pole;
C) the same as the weight of the pole.

Concepts Class 12: Fluids

1 Atmospheric pressure is approximately 100,000 N/m^2. The total force exerted on the front surface of your body by the atmosphere is roughly:
A) 0; B) 1 lb; C) 100 lb; D) 1000 lb; E) 10,000 lb. Explain.

Demonstration Clicker Question

2 A sheet of paper with dimensions 0.2 m x 0.25 m rests on a desk. What is the total force exerted by the air on the top surface of the paper? A) 500 N upward; B) 500 N downward; C) 5000 N upward; D) 5000 N downward; E) 0.
Explain. How could one demonstrate the reality of this force?
(Demonstrate large suction cup lifting heavy object.)

3 Is the pressure greater at a point A, 20 cm below the surface of a large lake or at point B, 20 cm below the top surface of milk in an open container? (Milk has very nearly the same density as water.) A) greater at A; B) greater at B; C) same at A & B.

4 A wading pool is 0.5 m deep and contains 1000 gallons of water. Water can be drained from the pool through a 1 cm diameter plastic pipe near the bottom of the pool. Suppose you try to stop the flow of water from the pool by applying pressure with your thumb to the outside end of the open drain pipe. Would this require more force or less force than stopping with your thumb the flow of water from the bottom of a 10 m long, 1 cm diameter, open-ended vertical pipe filled with less than 1 gallon of water? A) more force for pool drain; B) less force for pool drain; C) same force.

5 Two glass containers with different shapes, but the same size bases, are filled with water to the same height. In which of the containers does the water exert a greater force on base, or are the forces on the bases the same (C)?

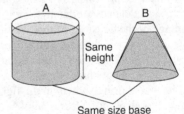

6 A submerged submarine, which weighs 3,000,000 N, is neither rising nor falling. What is the buoyant force on the submarine? A) < 3,000,000 N; B) 3,000,000 N; C) > 3,000,000 N; D) 0.

7 Is there a significant difference between the atmospheric pressure on the top and bottom of the helium balloon? A) yes; B) no.

8 This diver is at rest suspended below the surface of the ocean, but above the bottom. What happens if she takes a deep breath of air from the compressed air stored in a tank on her back?
A) she goes up to the surface;
B) she goes down to the bottom;
C) she doesn't move.

9 A mixed drink consists of a mixture of alcohol and water. The density of the mixture is a weighted average of the densities of the two liquids; that is, the overall density is the sum of the density of each liquid times the fraction of that liquid in the mixture. What can you conclude about the alcoholic content of the mixed drink shown in the figure? A) more than 50% alcohol; B) less than 50% alcohol.

$\rho_{alcohol} = 0.8 \rho_{water}$ $\rho_{ice} = 0.9 \rho_{water}$

Second Semester

Concepts Class 1: Coulomb's law, Electric Field

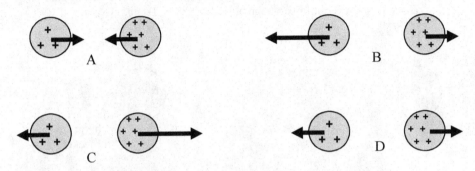

1 Two balls are positively charged and are close to each other, with the ball on the right side having more charge than the ball on the left. Which of these figures shows the correct magnitudes and directions of the forces on the balls?

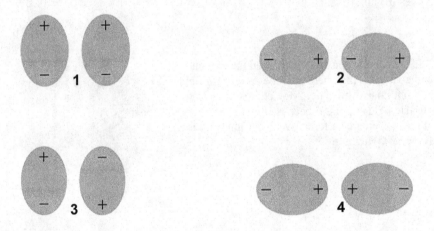

2 Water molecules have an electric dipole moment, meaning that one end is positive and the other is negative. For which of the pairs of dipoles in this figure is there a net attractive force between the dipoles? A) 1 only; B) 2 only; C) 3 only; D) 1 & 2; E) 2 & 3.

3 A small positively charged object is brought close to one end of a long metal rod that is electrically insulated and initially uncharged. The object does not touch the rod. Does the rod exert a force on the object? A) yes; B) no. Explain.

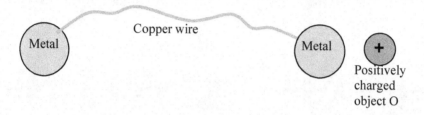

4 Two initially uncharged metal spheres are connected by a copper wire. A positively charged object O is placed near one of the spheres but not touching it. Is there any way that you can cause the two spheres to have a charge and to retain it even after the object is moved far away, without ever having contact with O? A) yes; B) no. Explain.

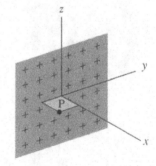

5 Joe observes a positively charged metal plate and concludes that there must be an electric field at point P near the plate. Moe claims there is no electric field at point P because there is no charge at P to experience a force. Who is right? A) Joe; B) Moe; C) neither.

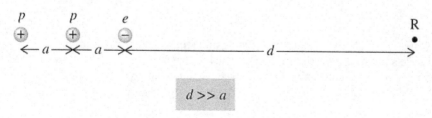

6 What is the direction of the electric field at point R, due to the electron and two protons shown in the figure above? A) left; B) right; C) up; D) down; E) there is no electric field at R – it's too far away.

7 In the old Star Trek TV series, prisoners on the Starship Enterprise were confined to their cabins by means of an invisible force field in an open doorway. When they attempted to pass through the doorway, they received a painful shock. Suppose the field is just a strong electric field. A prisoner reasons that since his body is uncharged, an electric field should not bother him. Is anything wrong with that reasoning? A) yes; B) no. Explain.

Concepts Class 2: Continuous Charge Distributions

1 Two uniformly charged infinite planes are shown in the figure. If a positively charged balloon of negligible weight is attached by a string to point P, what will be the direction of the string when the balloon comes to rest?

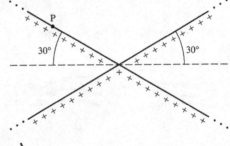

A) | B) —   C) /   D) \

2 What is the value of the electric field just outside the surface of a large, uniformly charged sheet of metal with surface charge density σ? A) $E = 2\pi k\sigma$; B) $E = 4\pi k\sigma$; C) $E = 0$.

Questions 3 – 8 You are hiking in the mountains during a thunderstorm. Would it be safe to:
3 stand directly under a tree? A) yes; B) no.

4 climb a tree? A) yes; B) no.

5 get inside an abandoned car? A) yes; B) no.

6 stand on the highest ground possible, a sharp peaked rock? A) yes; B) no.

7 seek shelter in a cave? A) yes; B) no.

8 seek shelter in a wooden house with a metal roof? A) yes; B) no.

9 At which of the points A, B, C, D is the field strongest? A; B; C; D; E (same at all points).

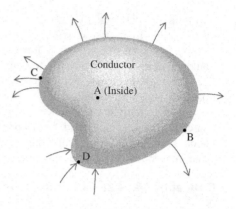

10 Which of the sketches below can represent electric field lines?

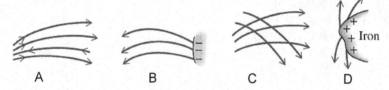

E) None of the sketches can represent electric field lines.

Concepts Class 3: Electric Potential

1 A + 3 C charge travels from the + terminal to the − terminal of a battery. How much electrical potential energy does it lose? A) 12 V; B) 36 V; C) 4 J; D) 36 J; E) − 36 J.

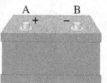

$V_A - V_B = 12 \text{ V}$

2 What can you conclude about the potential V on or in a conductor?
A) nothing, in general;
B) $V = 0$ throughout the conductor;
C) $V = 0$ on the surface;
D) V is constant on the surface, but not necessarily inside;
E) V is constant on the surface and inside.

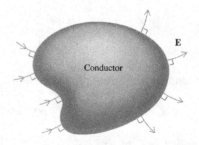

3 To calculate the potential difference between points P and R, use the equation $V_P - V_R = Ed$, where d equals: A) 5 cm; B) 4 cm; C) 3 cm; D) 0.

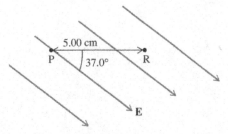

Demonstration Clicker Question

A Jacob's ladder is a dramatic device that was often used in old science fiction movies. A high voltage is applied to two long vertical wires, with a gap between them that increases from bottom to top. The voltage is sufficient to ionize the air, producing an arc of light between the two wires and a buzzing sound. The discharge begins at the bottom, where the distance between the wires is least and the electric field strongest. Then the heated air and arc rise until it disappears near the top, then starts over again at the bottom. Air ionizes when an electric field reaches approximately 3×10^6 N/C.

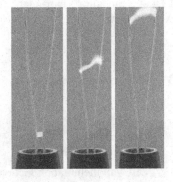

4 Observe the spacing between the wires and estimate the minimum voltage that must be applied to the two wires in the Jacob's ladder to produce the observed discharge. A) 3 V; B) 30 V; C) 300 V; D) 3,000 V; E) 30,000 V.

5 At which point, A, B, C, or D is the electric field directed toward the right?

6 At which point is the electric field greatest?

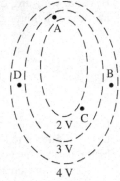

7 Charges q_1 and q_2 produce a potential V at a certain point P. If q_1 alone were present, the value of the potential at P would be $V_1 = 3$ V; if q_2 alone were present, the potential would be $V_2 = 4$ V. With both charges present, what is the potential at P? A) 3 V; B) 4 V; C) 7 V; D) 1 V; E) impossible to say.

8 At which point is the potential greatest?

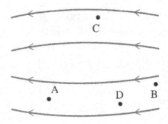

Concepts Class 4: Capacitance

1 Does a capacitor's capacitance depend on the voltage across its plates?
A) yes; B) no.

2 Does a capacitor's capacitance depend on the charge on its plates? A) yes; B) no.

3 Does a capacitor's capacitance depend on the size of its plates? A) yes; B) no.

4 Does a capacitor's capacitance depend on the separation between its plates?
A) yes; B) no.

5 A cell wall membrane functions as a capacitor: it is a dielectric and charge is stored on its inner and outer surfaces. What is the direction of the electric field inside the cell membrane? A) radially outward;
B) radially inward; C) there is no electric field inside the membrane.

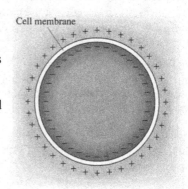

6 A person who has become electrostatically charged is about to touch a large metal plate with his fingertip. The positive charge on his finger induces an opposite charge on the plate. When the finger is a few millimeters away, there is a spark. Air becomes ionized and there is a flow of charge between finger and plate. The spark is painful, but not lethal. This is because which of the following is so small? A) voltage;
B) electric field; C) charge density; D) charge.

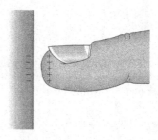

7 The two charged conductors in the figure serve as a capacitor. If Q increases, which of the following does *not* change? A) electric field; B) voltage; C) capacitance.

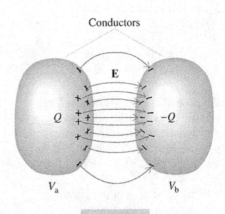

8 The two capacitors in the figure are connected in: A) series; B) parallel; C) neither series nor parallel.

Concepts Class 5: Electric Current, Ohm's Law

A proton moves to the right through this surface.
1 The current to the right is:
A) positive; B) negative.

2 The current to the left is:
A) positive; B) negative.

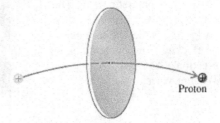

An electron moves to the left through this surface.
3 The current to the right is:
A) positive; B) negative.

4 The current to the left is:
A) positive; B) negative.

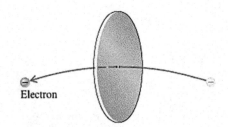
Electron

Because of the earth's atmospheric electric field, which is directed downward during clear weather, positive ions flow downward in the atmosphere and negative ions flow upward.
5 What is the sign of the current in the downward direction resulting from the flow of the positive charges? A) + ; B) – .
6 What is the sign of the current in the upward direction resulting from the flow of the negative charges? A) + ; B) –.

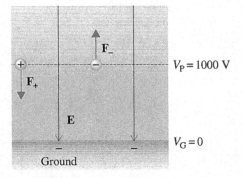

7 You are hiking in the mountains during a thunderstorm. Which of the following would be safest? A) stretch your body out flat – spread eagle – on flat, open ground; B) crouch down with feet close together on flat, open ground. HINT: When lightning strikes, it spreads out along the ground. Cattle are often killed by lightning while standing in open fields. Explain.

8 The glowing circle of light in this photo shows the path of electrons moving in a magnetic field. The electrons move around the circle clockwise. Is the current positive in the clockwise or counterclockwise direction? A) clockwise; B) counterclockwise.

Clicker Questions

9 Electrical potential energy is lost as an electric current passes through a resistor. Does this loss mean that the current leaving the resistor is less than the current entering?   A) yes; B) no.

Demonstration Clicker Question

This photo shows a mechanical system that models the motion of conduction electrons through a metal by marbles that collide with nails as they roll down an inclined surface. The conduction electrons are represented by the marbles.

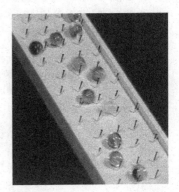

10 What does the vertical drop represent?
A) voltage; B) current; C) resistance.
11 What do the nails represent? A) voltage;
B) current; C) resistance.
12 The conduction electrons: A) gain kinetic energy and lose potential energy; B) lose potential energy, but kinetic energy does not change.

Concepts Class 6: DC Circuits

Demonstration Clicker Question

1 In these photos, a frayed electrical cord causes wires to touch, creating a short circuit when plugged into an electrical outlet. Why? A) a large voltage is applied to a very small resistance;
B) heat causes the current; C) there's no place else for the current to go.

 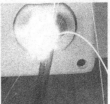

2 The current at point b is: A) + 12 A up; B) + 2 A up; C) + 6 A down; D) + 2 A down; E) − 4 A up

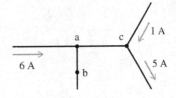

Questions 3 – 6: The resistors in this circuit represent light bulbs.

3 Which switches must be open in order to have an open circuit?
A) only S_1; B) only S_2; C) only S_3; D) all switches.

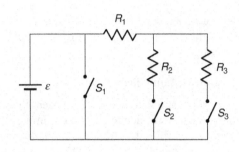

4 Which switches must be closed in order to have a short circuit? A) only S_1; B) only S_2; C) only S_3; D) all switches.

5 Is it possible to light bulb 2 without lighting bulb 3? A) yes; B) no. If so, how?

6 Is it possible to light bulb 1 without lighting bulb either bulb 2 or bulb 3? A) yes; B) no. If so, how?

7 Which resistors, if any, are in parallel? A) 1 Ω & 3 Ω; B) 1 Ω & 4 Ω; C) 2 Ω & 3 Ω; D) 2 Ω & 4 Ω; E) none.

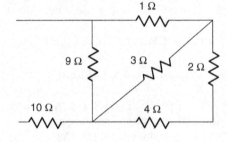

8 Which resistors, if any, are in series? A) 1 Ω & 3 Ω; B) 1 Ω & 4 Ω; C) 2 Ω & 3 Ω; D) 2 Ω & 4 Ω; E) none.

9 Identical resistors R_1, R_2, R_3, and R_4 carry currents I_1, I_2, I_3, and I_4 respectively. What are the relative magnitudes of these currents?
A) $I_1 = I_2 > I_3 = I_4$; B) $I_1 = I_2 > I_3 > I_4$; C) $I_1 > I_2 > I_3 > I_4$; D) $I_3 = I_4 > I_1 = I_2$; E) $I_1 = I_2 = I_3 = I_4$

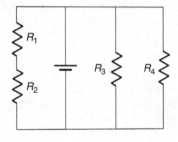

10 Resistors R_1 and R_2 are identical light bulbs. If a wire is connected between points P and Q, what happens to the brightness of the bulbs?
A) both stay the same; B) R_1 gets brighter and R_2 goes out; C) R_2 gets brighter and R_1 goes out; C) both get dimmer; D) R_1 gets dimmer and R_2 gets brighter.

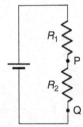

11 Are household electrical receptacles connected in series or parallel? A) series; B) parallel. Why?

12 A small insulated metal sphere carries a charge of 10^{-7} C and is at a potential of 10,000 V. Your left hand is grounded. Would it be more dangerous for your right hand to touch the sphere or to touch a power line at 120 V? A) touching the sphere is more dangerous; B) touching the power line is more dangerous. Explain.

Concepts Class 7: Magnetism

1 Positive and negative charge are located at the points shown in the figure. A uniform magnetic field is directed along the positive y-axis. The direction of each charge's velocity vector is indicated in the figure. Find the direction of the instantaneous force, if any, on each charge, a, b, and c. A) no force; B) $+x$ direction; C) $+y$ direction; D) $+z$ direction; E) in the x-z plane, 45° left of $+z$ axis;

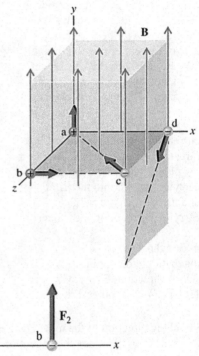

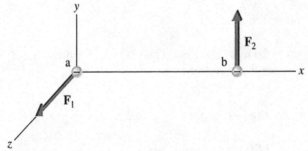

Because they are moving through a uniform magnetic field, the electrons at points a and b in the figure experience the instantaneous forces indicated.
2 What are the possible directions of the magnetic field? A) $+x$ direction only; B) $+$ or $-x$ directions; C) $+y$ direction only; D) $+$ or $-y$ directions; E) $+$ or $-z$ directions.

3 Consider the design shown in the figure. A circular metal ring has a metal center post, and a rigid wire that makes contact with the center post and the ring, but is free to rotate, so that the outer end slides over the ring. The center post and ring are connected to a battery as shown, so that current is provided. When placed in an external magnetic field, will the wire move?
A) no; B) yes, briefly to the right; C) yes, briefly to the left; D) yes, continuously, clockwise; E) yes, continuously, counterclockwise.

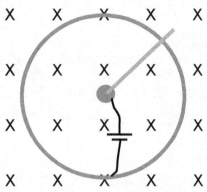

Two long straight wires carry currents of equal magnitude directed perpendicular to the x-y plane, the wire on the left carrying current into the page and the wire on the right carrying current out of the page.

4 Find the direction of the magnetic field at point P.
A) $+x$ direction; B) $+y$; C) $-x$; D) $-y$; E) some other direction.

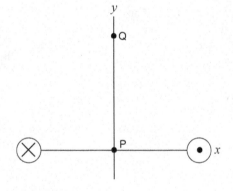

5 Find the direction of the magnetic field at point Q. A) $+x$ direction; B) $+y$; C) $-x$; D) $-y$; E) some other direction.

Concepts Class 8: Waves; Light

An electrocardiogram (ECG) is a graph showing electrical oscillations resulting from the beating of a human heart.
1 Is such motion approximately harmonic? A) yes; B) no.
2 Is such motion approximately simple harmonic? A) yes; B) no.

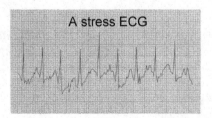

A stress ECG

3 A child swings back and forth on a playground swing, with an amplitude of 10°. Suppose you increase the amplitude of the motion from 10° to 20°. How does the child's average speed change? A) it doesn't change; B) it doubles; C) it more than doubles; D) it decreases; E) none of the above.

4 Traffic is stalled on a freeway. When cars finally begin to move, the motion passes through the line as a wave pulse. What is the direction of the pulse relative to the motion of the cars? A) same direction; B) opposite direction.

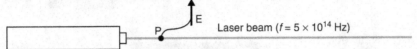

5 At point P, at time t, the electric field of the laser beam is $\mathbf{E} = \mathbf{E_0}$ = 1000 N/C, directed up. What is the electric field at P at a time 10^{-15} s later?
A) 1000 N/C, up; B) 1000 N/C, down; C) 0; D) 500 N/C, up; E) 500 N/C, down.

6 HeNe laser light, of wavelength 630 nm in air, shines into a body of water. What is the wavelength of the light in water? A) 630 nm; B) < 630 nm; C) > 630 nm.

7 What color is the light underwater? A) the same as out of water; B) shifted toward the red; C) shifted toward the blue.

8 With the line of sight directed at point P, which of the points in the figure could the observer see: A, B, C, D, or E?

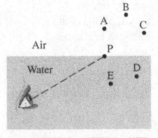

9 Which of the points A through E could be the source for the light ray shown in the figure? A) only A; B) only C; C) only E; D) A or E; E) C or E.

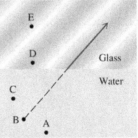

10 A light ray is initially in glass and strikes the surface that the glass makes with air at an angle of incidence of 60°. At what angle does the light emerge from the glass? A) 35°; B) 60°; C) some other angle; D) the light does not escape.

Concepts Class 9: Mirrors, Lenses

1 Which is the mirror image? A) image on left; B) image on right. Explain.

2 This photo shows that there are three images of an object placed in front of two perpendicular plane mirrors. How many images are formed by two mirrors at an angle of 60°, as seen in the figure to the right? A) 2; B) 3; C) 4 or more.

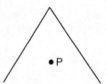

3 Draw ray diagrams to determine the number and location of all the images of the point P in front of the two mirrors.

4 You see the image of a person in a mirror. Will the person necessarily be able to see you by looking in the mirror if there is sufficient illumination? A) yes; B) no; C) it depends. Explain.

5 Which of these figures shows the correct path of a light ray reflected from a spherical mirror?

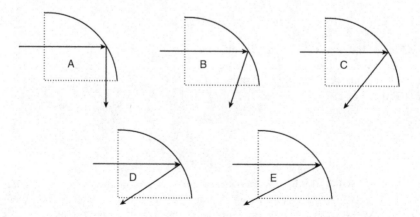

6 A laser beam is incident on a spherical drop of water. Which figure shows the correct path of the light through the water?

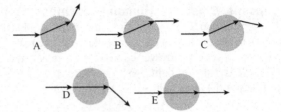

7 A laser beam is incident on a spherical drop of water. Which figure shows the correct path of the light through the water?

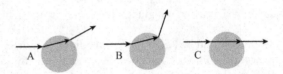

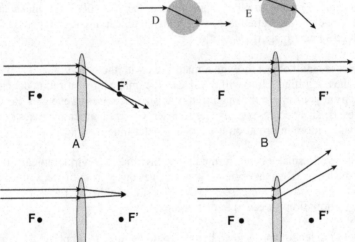

8 Horizontal light rays are incident on a positive lens. Which of the figures above shows the correct paths of light rays after they leave the lens?

9 Parallel light rays, directed below the horizontal, are incident on a positive lens. Where will these rays converge to a point? A) at the focal point; B) to the right of the focal point; C) to the left of the focal point; D) directly below the focal point.

Concepts Class 10: Human Eye, Optical Instruments

1 Squinting can allow a nearsighted person who is not wearing glasses or contact lenses to see distant objects more clearly. The reason for this is that: A) less light enters the eye; B) narrower cones of light rays enter the eye; C) pressure is exerted on the lens, causing light rays to be better focused on the retina; D) there is more diffraction; E) there is less diffraction.

2 Suppose you are nearsighted and are trying to read distant road signs without glasses or contact lenses to correct your vision. This is: A) easier in bright daylight; B) easier at night; C) the same day or night.

3 Which would be easier to resolve with the normal, unaided eye under favorable viewing conditions? A) 2 people standing side by side 1 m apart at a distance of 1000 m; B) 2 stars, 0.1 light years apart, each of which is 10 light years from earth.

4 Magnifiers, microscopes, and telescopes all have one thing in common: their function is to: A) make an object appear closer than it really is; B) make an object's image on your retina larger; C) bring an object into sharp focus; D) produce a linear magnification > 1.

5 You view a small object through a magnifier, with the object placed at the focal point of the magnifier. How will the size of the image on your retina depend on the distance from your eye to the magnifier? A) size increases as eye gets closer to magnifier; B) size decreases as eye gets closer; C) size neither increases nor decreases; D) more information is needed to determine.

6 The image of a small object, formed by a magnifier, is a virtual image 20 cm in front of the lens. When an observer's eye moves away from the lens, the size of the image on the retina: A) increases; B) decreases; C) neither increases nor decreases; D) more information is needed to determine.

7 Which of the lens shapes shown below would be most appropriate for contact lenses correcting nearsightedness? Explain.

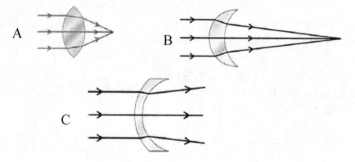

8 Suppose that a magnifier made of glass is replaced by a magnifier made of diamond ($n = 2.41$) with exactly the same shape as the glass magnifier. Compared with the glass lens, the size of the largest retinal image that can be produced using the diamond lens is: A) larger; B) smaller; C) the same. Explain.

Concepts Class 11: Interference

1 Knowing the phase of a wave at an initial instant, and the frequency, can we predict the phase a time Δt later? Suppose, for example, the light is from a sodium source ($f = 5 \times 10^{14}$ Hz), & $\Delta t = 10^{-3}$ s. If the wave is at a peak at time t_0, will it be at a peak at that point a time $\Delta t = 10^{-3}$ s later? A) yes; B) no; C) Impossible to say.

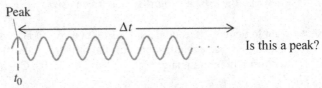

2 Is common He Ne laser light coherent over 10^{-9} s?
A) yes; B) no.

3 Is white light coherent over 10^{-9} s? A) yes; B) no.

Light Source	Frequency (Hz)	Bandwidth (Hz)
White light	$4 \times 10^{14} - 7 \times 10^{14}$	10^{14}
Sodium line	5×10^{14}	10^{9}
HeNe laser	4.7×10^{14}	10^{8}
Stabilized laser		10^{4}

4 Which of the following light sources above is coherent over a distance of 0.2 m? A) common HeNe laser; B) sodium vapor lamp; C) white light; D) none.

5 Suppose that Young's double-slit experiment is performed underwater. The interference fringes would be: A) closer; B) farther apart; C) spaced the same. Explain.

6 Suppose you trap a layer of water 1 mm thick between two microscope slides (glass plates) and look directly down at the layer. You would see: A) no light due to destructive interference caused by reflection from the two surfaces; B) enhanced light due to constructive interference caused by reflection from the two surfaces; C) colors corresponding to the wavelengths present in the illuminating light being either constructively or destructively interfering; D) no interference in the reflected light.

7 This soap bubble is colored because of interference of light reflected from its surfaces. Just before a bubble bursts, its thickness is much less than the wavelength of light. What color will the bubble then appear? A) red; B) blue; C) yellow; D) white; E) black.

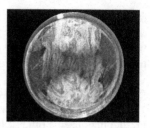

Concepts Class 12: Diffraction and Polarization

1 For a given diameter objective lens in a microscope, what color light will produce the smallest diffraction circle as an image of a point on the object? A) red; B) yellow; C) green; D) blue; E) all colors give the same size.

2 Suppose we want to demonstrate single-slit diffraction over a large angle. In principle, if we had an intense enough source of light, the central diffraction maximum could cover a maximum angle of: A) < 1°; B) < 10°; C) 45°; D) 90°; E) 180°.

3 We want to demonstrate diffraction by a thin slit over a very large angle. But light sources are not intense enough to be seen when the slit is narrow enough to produce very wide angle diffraction. To see wide angle diffraction, we can: A) use mirrors to reflect and concentrate the light source; B) use lenses to concentrate the light source; C) put lots of very tiny slits close together; D) do nothing; we just can't see wide angle diffraction.

4 Two Polaroid sheets are oriented with perpendicular transmission axes, as shown. Unpolarized light is incident from the left side. **a)** Will any light be transmitted through the sheet on the right? A) yes; B) no. **b)** Suppose a third Polaroid sheet is placed between the two sheets shown here. Could any light possibly be transmitted through the sheets? A) yes; B) no.

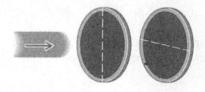

5 The lenses in these sunglasses have a Polaroid sheet. The transmission axis should be: A) vertical; B) horizontal.

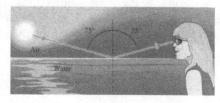

Concepts Class 13: Relativity

1 Your precise time of birth was _____ (hour, minute, & date). That time was determined *solely* by: A) the absolute time of your birth; B) the absolute time *on earth* of your birth; C) the absolute time and place on earth of your birth; D) the reading of clocks simultaneous with your birth.

2 Is there any scientific meaning of time of events on earth, other than the readings of earth clocks? A) yes, because there may be no nearby clocks for some events; B) no, the only meaning of time is the reading of a clock.

3 In the movie *Contact* mankind's first communication from an extraterrestrial civilization comes from the star Vega, 27 light years from earth, begins with a re-broadcast of one of the earliest television transmissions on earth, showing Adolf Hitler addressing spectators at the 1936 Olympic Games. What is the earliest year that a re-broadcast of this signal from Vega could have been received on earth? A) 1936; B) 1963; C) 1990; D) 1997.

4 The earliest year that the signal could have left Vega is: A) 1936; B) 1963; C) 1990.

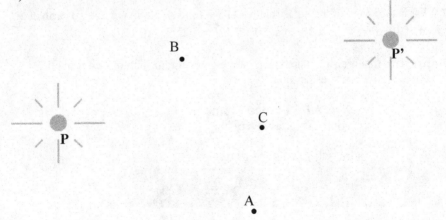

5 Observers at points A, B, and C witness the explosion of supernovas at points P and P'. In order for the supernova explosions to be simultaneous, they must be seen simultaneously at point: A; B; C.

6 A laser pulse is emitted from a spaceship moving at 2×10^8 m/s relative to earth. What is the speed of the pulse relative to the earth and moon? A) 3×10^8 m/s; B) 4×10^8 m/s; C) 5×10^8 m/s.

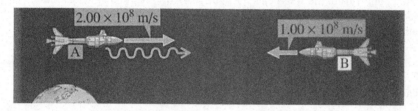

7 What is the speed of the pulse relative to a second ship moving toward earth at a speed of 10^8 m/s? A) 3×10^8 m/s; B) 4×10^8 m/s; C) 5×10^8 m/s.

8 Are events that are simultaneous in one reference frame simultaneous in all reference frames? A) yes; B) no.

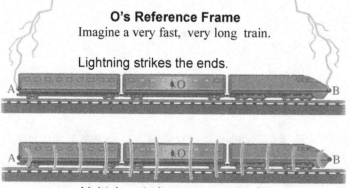

9 Lightning strikes both ends of a train that is moving in a straight line at a speed comparable to the speed of light. An observer O, riding on the train and located at its midpoint, sees the flashes of light simultaneously. An observer on the ground would conclude that: A) lightning strikes both ends simultaneously; B) lightning strikes the back end of the train first; C) lightning strikes the front end of the train first.

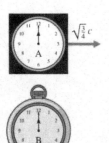

10 Clock A travels to the right at a speed of $(3/4)^{1/2}c$ relative to clocks B and C. As A passes B, all clocks read 12:00. These readings could be seen simultaneously by a stationary observer O, midway between B and C, though O would make the observation at some later time. Clock C reads 6:00 as A passes C. What does A read? A) 3:00; B) 6:00; C) 9:00; D) 12:00.

Concepts Class 14: Quantum Mechanics

1 Our best description of the nature of light in a laser beam is 6×10^{16} particles of light traveling through the laser beam every second. A) true; B) false.

2 Do all of the electrons in a multielectron atom go into the lowest possible energy electron state ($n = 1, l = 0, m = 0$)? A) yes; B) no.

3 In the double-slit experiment can we think of each photon of light as traveling either through one slit or the other? A) yes; B) no.

4 Helium was discovered on the sun in 1868, 27 years before it was found on earth. Which of helium's properties allowed for its discovery in this way? A) chemical interaction with other elements; B) wavelengths of spectral lines; C) atomic weight; D) speed of sound through helium.

Problem Solving Worksheets

"The formulation of the problem is often more essential than its solution, which may be merely a matter of mathematical or experimental skill."   Albert Einstein

Worksheet A   Problem # _____   Name_____
I Formulate the Problem
Sketch

Key Facts (e.g. frictionless):

Data:
Express problem in terms of symbols:

II Plan What kind of problem is this? Check one or more.
Vector addition ☐
Kinematics ☐
Dynamics ☐
Center of mass ☐
Universal gravitation ☐
Momentum ☐
Energy ☐
Static equilibrium ☐

Problem Solving Worksheets

Kinematics Worksheet B   Problem # _____   Name_____

II Plan What kind of kinematics problem is this? Check one or more.
Linear motion ☐
Two dimensional motion ☐
Constant acceleration ☐
Free fall ☐
Projectile motion ☐
Circular motion ☐
Relative motion ☐

Which general equations for this type of motion can I use to solve this problem?

How will I use these equations?

III Execute Plan Work through the math

IV Review
Does the solution make sense?
Do units check?

Dynamics Worksheet B   Problem # ____   Name_____

II Plan Choose a body as your free body. Draw a circle around this body in your sketch on worksheet A. (If necessary, choose more than one free body.)
Is the body accelerated?
Free-Body Diagram(s) Draw external forces. Show a coordinate system.

```
┌─────────────────────────────────────────────────────────────────┐
│                                                                 │
│                                                                 │
│                                                                 │
│                                                                 │
│                                                                 │
│                                                                 │
└─────────────────────────────────────────────────────────────────┘
```

Write Newton's 2nd law in component form (2 equations for 2 dimensions)

Force laws that apply (weight on earth, tension, spring force, normal force, friction, universal gravitation, buoyant force):

How will I use these equations?

III Execute Plan Work through the math

IV Review
Does the solution make sense?
Do units check?

Problem Solving Worksheets

Energy Worksheet B   Problem # ____   Name_____

II Plan Choose a body as your free body. . Draw a circle around this body in your sketch on worksheet A.

Free-Body Diagram

```



```

Are any nonconservative forces acting on the free body?
If there are nonconservative forces, do they do work?
Is mechanical energy conserved?   If so, write this general principle, and write out the general equation relating the initial and final energies of all types.
Indicate in your sketch on worksheet A the initial and final states.
If mechanical energy is not conserved, write another appropriate equation you could use.

How will I use these equations?

III Execute Plan Work through the math

IV Review
Does the solution make sense?
Do units check?

12

Labs and Lab Concept Quizzes

"We want the facts to fit the preconceptions. When they don't it is easier to ignore the facts than to change the preconceptions." Jessamyn West

FIRST SEMESTER ONLY (2nd semester labs available on TIP website)
Lab 1: Variables, Motion
Equipment: variable speed model car, metronome, paper, scotch tape, ruler, tape measure, video camera and cable, tripod, LoggerPro software, computer
I Variables
In the figure below, identify variables and values of variables.

	Value 1	Value 2	Value 3
Variable 1:			
Variable 2:			
Variable 3:			

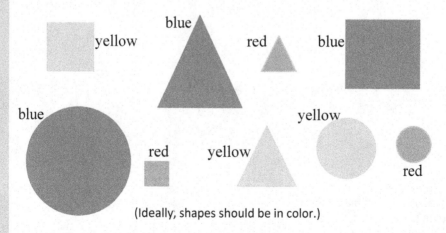

(Ideally, shapes should be in color.)

Is there a relationship between the variables? That is, are certain values of one variable related to certain values of another variable, so that when one variable changes, the other also changes? For each combination of variables in the table below, indicate by a check in the appropriate column whether the variables are related or unrelated.

		Related	Unrelated
Variables 1 & 2:	&		
Variables 1 & 3:	&		
Variables 2 & 3:	&		

Give some examples of variables and their values for human beings.

Human Variables	Some Values of the Variables

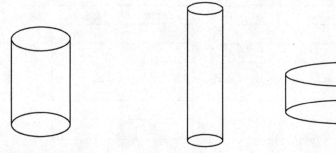

Suppose the same amount of water was placed in each of the three containers shown here. What variables could you use to describe the water in the containers? Are these variables related in any way?

II Describing the Motion of a Motorized Model Car

1 Turn on the car, adjust it to the minimum, and place it on the tabletop so that it travels in a straight line. Name 3 variables you could use to describe its motion.

Now let's quantify your description of the car's motion by measuring its position x corresponding to time t. Make multiple measurements of those variables for a single trip. A metronome set at 60 beats per minute will emit a sound every 1.0 s. Tape paper to the tabletop so that you can make marks on it as the car travels over the paper. Describe precisely how you make your measurements. (Turn the volume down on your metronome and hold it close to your ear so that others' metronomes don't interfere.) Sketch the trail of the car. Use dots, labeled 0,1,2,... Record your measurements of the two variables x and t in a table.

$\longrightarrow x$

2 Repeat **1**, but with the car's motor turned up significantly (but not so much that measurement is difficult).

$\longrightarrow x$

1

x, cm	t, s

2

x, cm	t, s

Graph x vs. t for **1** and **2** on a single graph. Use labels 1 & 2 to distinguish the two runs.

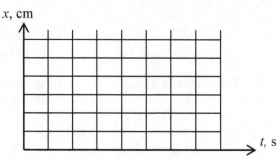

3 Velocity v_x, derived from the variables x and t, is another variable used to describe motion. It is defined as the **rate of change of position x.** Use either your tables or your graph to determine values of velocity for the two cars.

1 $v_x =$ _____   **2** $v_x =$ _____

Graph v_x vs. t for the two data sets on the graph below.

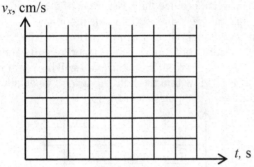

Suppose the motion of the car were to continue indefinitely. Is there a way you could predict the position of the car at a later time, say 30 s later? Explain.

4 Sketch the trail and graphs of x vs. t and v_x vs. t for a real car that first moves at a steady speed, then slows to a stop at a traffic light, then, a few seconds later, starts up and returns to its initial speed. Make up reasonable numbers.

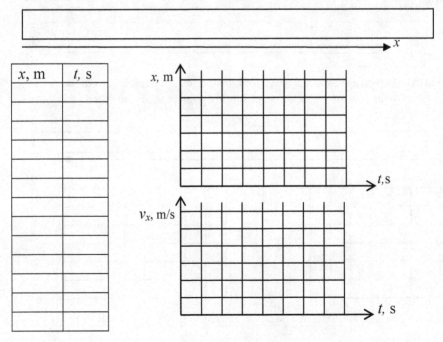

III Human Motion Experiments

You will explore the linear motion of you and your partner. In each part, you will be given one of the following: a description in words, a series of dots showing a sequence of equal-time positions, or a graph of either position or velocity vs. time. In all cases, provide the other ways of describing the motions. For example, if a description in words is given, draw the other three ways of representing the motion. Or if a position vs. time graph is given, provide the description of the motion in words, along with the other graphical ways of describing the motion.

After you have made your predictions, check them by performing the experiment, using the camera to record your motion and LoggerPro software to graph data on the computer. Set up an x axis along which you will move. It should be perpendicular to the camera and well within the field of view. See detailed instructions on the last page for use of computer and camera.

1

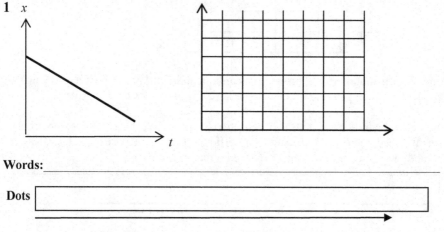

Words:_____

Dots

2 Words: Move at a slow steady pace from the origin to a point 2 m from the origin and then immediately move back to your starting point at twice the speed you were moving away.

Dots

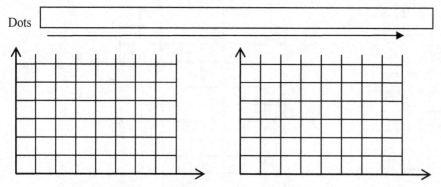

3

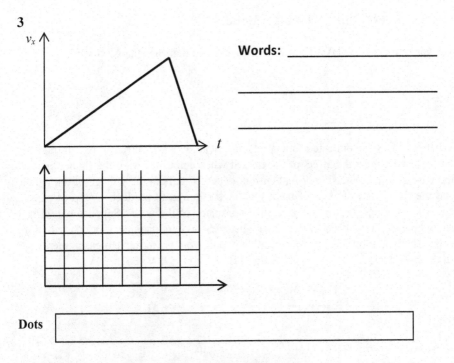

Words: _____

_____

_____

Dots [   ]

Recording Digital Video and Using LoggerPro

Turn on the video camera, which should be connected to a computer, and adjust the camera's position and field of view for the experiment you are performing. Place a meter stick in the background so that it is clearly visible in the camera field and is as close as possible to the object whose motion is to be measured, in order to minimize parallax. Click on the LoggerPro icon on your computer screen. Then click on Insert video capture. The camera's field of view should appear on the screen. Click Start just before the motion begins and click Stop just after it ends. Click on the lower right part of the video screen to bring up LoggerPro icons on the right side of that screen. Click on the meter stick icon and use the cursor to measure the meter stick, so that the computer has a length scale for its analysis. Then click on the crosshairs icon to allow you to record equal time images of an object in your movie. Make sure you advance the movie to the point where the moving object begins to move or to the part of the motion you want to analyze. Choose one point on the body to click on. For the human motion experiment you might choose the top of the shoulder. As you click, the image will advance frame by frame, giving you a dot sequence that shows the position of the object at equal time intervals. Graphs of x vs. t and y vs. t are generated automatically. Under the Analyze menu, click on either linear or curve to do a fit to your data points. To change from the x and y graphs, click on the x, y graph labels, and you can choose to graph v_x or v_y.

Lab 1 Concept Quiz: Variables, Motion

1 A variable is: A) x; B) y; C) x or y; D) something that can change or vary.

2

When a battery is connected to a conducting wire, electric current flows through the wire. You can feel the effect of the current when connected as in the figure, because the wire gets hot. The size of that current depends on both the voltage of the battery and the wire. The amount of battery voltage required to produce 1 amp of electric current in a wire is called the electrical resistance of the wire. Suppose we wish to study how electrical resistance of a wire depends on wire diameter. Which two of the copper and aluminum wires shown below would you use to make measurements and study this effect? A) I & II; B) I & III; C) II & IV; D) I & V; E) II & IV.
Explain.

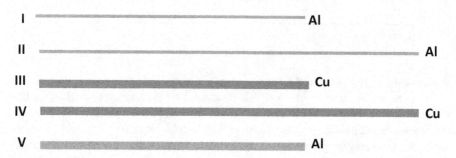

Constant Sum If adding a certain amount to one variable causes the same amount to be subtracted from the other variable, then the variables are related by a constant sum.
Constant Difference If adding a certain amount to one variable causes the same amount to be added to the other variable, then the variables are related by a constant difference.
Constant Ratio or Proportionality If there is a correspondence of a fixed amount of one variable for every certain amount of the other variable, then the variables are related by a constant ratio. If doubling one variable causes the other variable to double, then the variables are related by a constant ratio. The variables are proportional.
Constant Product or Inverse Proportionality If changing the value of one variable causes the other variable to change in such a way that the product of the two variables is constant, then the variables are related by a constant product. If doubling one variable causes the other variable to be halved, then the variables are related by a constant product. The variables are inversely proportional.

Labs and Lab Concept Quizzes

For questions **3 – 15**, indicate whether the variables are related by: A) constant sum; B) constant difference; C) constant ratio; D) constant product.

3 A cook is roasting a turkey. The roasting time is 15 minutes per pound of turkey. The cooking time and the weight of the turkey are related by: ___

4 A student is doing an experiment flipping a coin 100 times. The number of heads and number of tails are related by: ___

5 Two cars are traveling towards LA at the same speed. The first car left earlier, so it is closer to LA. When the first car is 100 miles away, the second car is 150 miles away. The distance of the first car from LA and the distance of the second car from LA are related by: ___

6 On one particular map a distance of 2 inches represents an actual distance of 4 miles. The map distance and the actual distance are related by: ___

7 At a certain time of day a 6 foot tree cast a shadow 10 feet long. The height of the tree and the length of the shadow are related by: ___

8 A basketball team plays 30 games during the season. The number of wins and the number of losses are related by: ___

9 Two sisters, Jane and Mary, each have a savings account. They both deposit $20 each month in their savings. Jane opened her savings account first, so when Jane has $100 in savings, Mary has $60 in savings. The amount of money in Jane's account and the amount of money in Mary's account are related by: ___

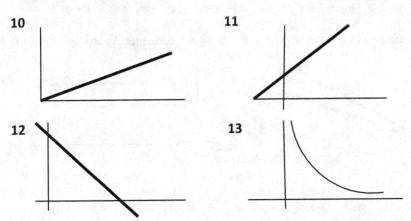

14 A car is travelling at a constant speed of 50 miles per hour along a straight road. The relationship between the time and the distance traveled during that time are related by: ___

15 In a 100 m dash all the runners travel the same distance in different times. The runners' speeds and their times are related by: ___

16 – 27

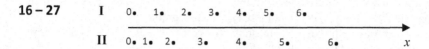

The two series of dots above indicate the positions of two moving bodies, I & II, at times 0,1,2,3,... The times could be in any units – seconds or hours for example. The time interval between successive dots is the same. The coordinate x measures the position. Indicate for each time, 0 to 5, whether object II's x-coordinate is: A) greater than, B) less than, or C) equal to I's x-coordinate at that time.

16 0 ___ **17** 1 ___ **18** 2 ___ **19** 3 ___ **20** 4 ___ **21** 5 ___

Indicate for each time interval, 0 to 1, 1 to 2, etc., in the figure above, whether object II's speed is A) greater than, B) less than, or C) equal to I's speed for that interval.

22 0 to 1 ___ **23** 1 to 2 ___ **24** 2 to 3 ___ **25** 3 to 4 ___ **26** 4 to 5 ___
27 5 to 6 ___

28, 29

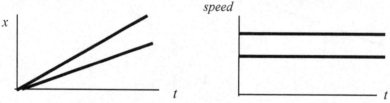

The dot sequences above and the graphs below represents the motion of the two bodies I & II.

28 For the x vs. t graphs, the line with the steeper slope represents: A) body I; B) body II.

29 For the speed vs. t graphs, the line that is higher represents: A) body I; B) body II.

30 You and your partner are part of a top secret team watching out for UFO's that have been sighted, but don't show up on radar. After long months of observation, your patience pays off one night when you see a UFO directly over the LMU campus and exactly 10 seconds later your partner sees the same UFO directly overhead at her location, which is the Hollywood sign, approximately 20 km from LMU, north. You want to be able to tell the air force where the UFO will be 15 minutes later (the length of time it will take to get a plane to pursue). Where would you tell them to look, measured from LMU? A) 30 km north; B) 50 km north; C) 1200 km north; D) 1800 km north; E) 500 km east. Discuss any assumptions you make.

Lab 2: Acceleration; Relative Motion

Equipment: low friction Pasco model car, 2 m track, wooden blocks, metronome, chalk, tape measure, meter stick, tennis ball, video camera and cable, tripod, LoggerPro software (adjust settings so that 3 points, not 7, are used for derivatives), computer

I Linear Acceleration

1 Incline the track by placing a 1 cm block under the legs at one end. Release the car from the high end of the track and observe its motion. Using the metronome and chalk, mark its position at equal time intervals on the tabletop. In the box below use dots to show the position of the car at equal time intervals.

What can you conclude about the car's velocity as it rolls down the track?

2 You are in your car stopped at a traffic light. You step on the gas and begin moving forward, your speed steadily increasing for 30 s until you reach 60 mi/hr. Your speedometer reading increases 10 mi/hr every 5 s. Sketch a graph of v_x vs. t. It's OK here to use mixed units, mi/hr for velocity, s for time.

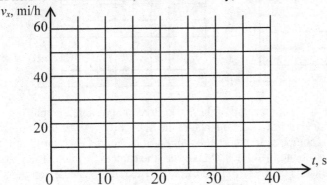

What is the value of the slope? _____ The slope is the **rate of change of velocity**, which we call **acceleration** and denote by a_x. Sketch graphs of a_x vs. t and x vs. t for the car. Just sketch the shape, no numbers.

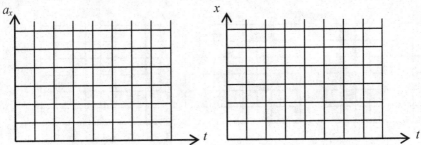

3 Continuing the description of motion in **2**, you now apply the brakes and your car slows to a stop, decreasing 20 mi/hr every 5 s, starting from an initial velocity of 60 mi/hr. Graph v_x vs. t, then sketch qualitative graphs of a_x vs. t and x vs. t.

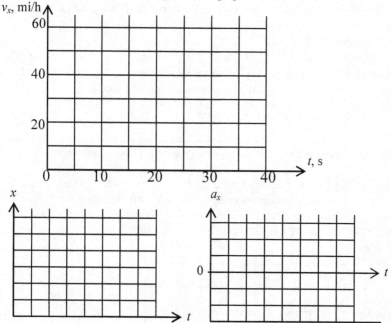

4 Raise the legs 7 or 8 cm on one end of the track. Predict and sketch graphs of $x, v_x,$ and a_x for a model car released from rest at the high end of the track. Then use the video camera and computer to record the motion of the model car, starting from rest and rolling down the inclined track. Make sure the camera is oriented so that the high end of the track is on the left in the camera's field of view. Click on the axes icon, set the origin at the starting point, and rotate the axes so that the car's motion is along the $+ x$ axis. Observe graphs of x, v_x, and a_x vs. t. Sketch the graphs you observe. Check results with your TA or instructor.

Predicted & Observed Sketches

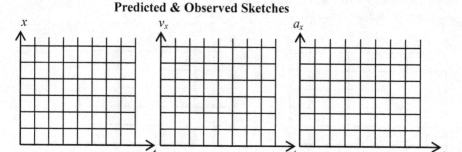

5 Now predict and sketch motion graphs for the car starting at the low end of the track with an initial velocity up the track. Then use the video camera and computer to record that motion. This time the low end of the track should be on the left in the camera's field of view, so that the positive x axis is up the track. Click on the axes icon, set the origin at the starting point, and rotate the axes so that the car's initial motion is along the $+x$ axis. Record the motion until the car rolls back down to its starting point. Observe graphs of x, v_x, and a_x vs. t. Sketch the graphs you observe. Check with your TA or instructor.

Predicted and Observed Sketches

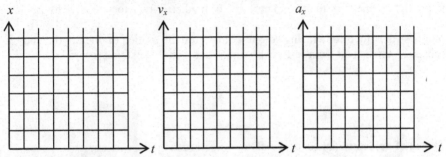

II Acceleration in 2-D

1 Predict and sketch motion graphs for the motion of a tennis ball that you toss a few meters across the room. Then use the video camera and computer to record the motion, and observe graphs of x, y, v_x, v_y, a_x, and a_y vs. t. Sketch the graphs you observe. Check with your TA or instructor.

Predicted and Observed Sketches

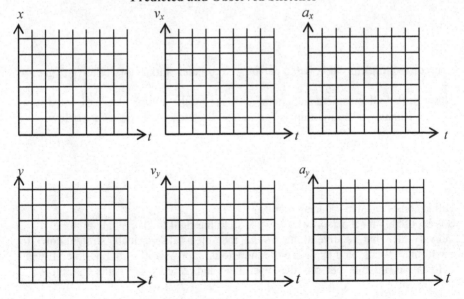

III Relative Motion

1 Consider the motion of a tennis ball released from rest by a person who is walking to the right at a steady pace. Predict and sketch the trajectory relative to: a) the walking person; b) the ground. Use a series of dots to indicate position at equal time intervals.

a) predicted trajectory relative to person   b) predicted trajectory relative to ground

Now use the video camera and computer to record the motion of one of you who releases a tennis ball from rest as she walks, and sketch what you record below.

Sketch of trajectory/dot sequence of ball as recorded by the camera.
Is this consistent with your prediction?

2 Consider the motion of a tennis ball thrown straight up by a person who is walking at a steady pace. Predict and sketch the trajectory/dot sequence relative to: a) the walking person; b) the ground.

a) predicted trajectory relative to person   b) predicted trajectory relative to ground

Now use the video camera and computer to record the motion of one of you who throws a tennis ball straight up as she walks, & sketch what you record below.

Is this consistent with your prediction?
3 Is it possible for a person who is walking to the right to give a ball an initial velocity such that, relative to the ground, the object falls vertically straight down to the floor? Discuss and explain either how to do it or why it is not possible. If you believe it is possible, perform the experiment and record your results with the camera.

Lab 2 Concept Quiz, Acceleration and Relative Motion

1 – 16 For each of the dot sequences and for each of the graphs, indicate whether the velocity v_x is: A) positive and constant; B) positive and increasing; C) positive and decreasing; D) negative and constant; E) negative and increasing; AB) negative and decreasing; AC) first positive and constant, then decreasing; AD) first negative and constant, then increasing.
In all cases, the positive x-axis is to the right. ⎯⎯⎯⎯⎯⎯⎯⎯⎯⎯⎯→ x

1   0. 1. 2. 3. 4. 5.6.   5   0. 1. 2. 3. 4. 5.6.

2   0. 1. 2. 3. 4.   5.   6.   6   6. 5. 4. 3. 2. 1. 0.

3   6. 5. 4. 3. 2.   1.   0.   7   6.   5.   4. 3. 2. 1. 0.

4   0. 1. 2. 3. 4. 5. 6.   8   6. 5. 4. 3.   2.   1.   0.

9
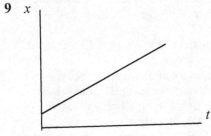

10
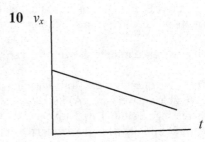

11
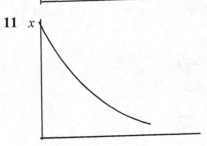

12
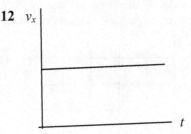

13
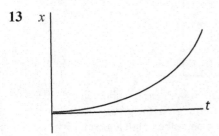

14
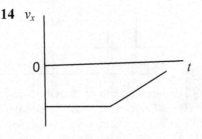

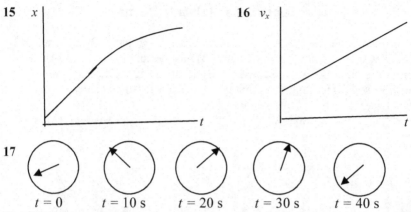

17 The figure above shows a car's speedometer at 10 s intervals. The car's acceleration in the forward direction is: A) positive throughout; B) positive for 10 s and then zero; C) positive for 10 s and then negative; D) positive for 20 s and then negative; E) positive for 30 s and then negative. Provide a story about the car's motion that corresponds to the speedometer readings.

Sketch a sequence of dots showing the car's position at t = 0, 10 s, 20 s, etc.

18 – 29 In the questions that follow asking about a component of velocity, either v_x or v_y, possible answers are: A) initially positive and constant; B) initially positive and increasing; C) initially positive and decreasing; D) initially negative and constant; E) initially negative and increasing; AB) initially negative and decreasing. In the questions that follow asking about a component of acceleration, either a_x or a_y, possible answers are: A) positive; B) negative; C) zero.

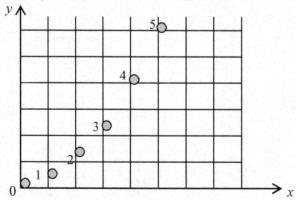

18 The x-component of velocity v_x for the motion in the graph above is:
19 The x-component of acceleration a_x for the motion in the graph above is:
20 The y-component of velocity v_y for the motion in the graph above is:
21 The y-component of acceleration a_y for the motion in the graph above is:

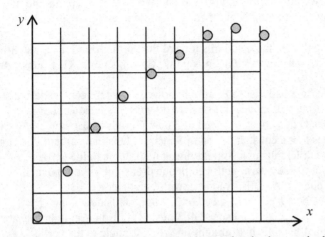

22 The x-component of velocity v_x for the motion in the graph above is:
23 The x-component of acceleration a_x for the motion in the graph above is:
24 The y-component of velocity v_y for the motion in the graph above is:
25 The y-component of acceleration a_y for the motion in the graph above is:

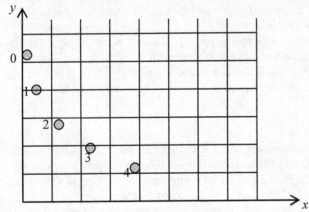

26 The x-component of velocity v_x for the motion in the graph above is:
27 The x-component of acceleration a_x for the motion in the graph above is:
28 The y-component of velocity v_y for the motion in the graph above is:
29 The y-component of acceleration a_y for the motion in the graph above is:
30 Is all motion relative, or is motion absolute, meaning that there is really only one right way to describe motion, though our perception of motion might be distorted? A) motion is relative; B) motion is absolute.
31 – 36 You are on a skateboard in the lab and you move through the lab at 3 m/s headed north as your partners sit at the table and observe you.

31 Relative to you, what is the motion of your partners? A) they are not moving; B) moving N at 3 m/s; C) moving N at 1.5 m/s; D) moving S at 3 m/s; E) moving S at 1.5 m/s.

32 A car moves along the lab table at 4 m/s toward the north. Relative to you, how does the car move? A) 4 m/s north; B) 3 m/s north; C) 2 m/s north; D) 1 m/s north; E) 1 m/s south.

33 Your partners release a ball that falls straight down to the tabletop. Relative to you, how does the falling ball move? A) straight down; B) falls down, but simultaneously moves N at a constant speed of 3 m/s; C) falls down, but simultaneously moves S at a constant speed of 3 m/s; D) falls in a straight line at a 45°angle to the vertical; E) falls in a straight line at a 30°angle to the vertical.

34 Still moving on your skateboard, your cell phone drops out of your pocket. Relative to you, the phone: A) falls straight down; B) falls down, but simultaneously moves N at a constant speed of 3 m/s; C) falls down, but simultaneously moves S at a constant speed of 3 m/s; D) falls in a straight line at a 45° angle to the vertical; E) falls in a straight line at a 30°angle to the vertical.

35 Relative to your partners, the cell phone: A) falls straight down; B) falls down, but simultaneously moves N at a constant speed of 3 m/s; C) falls down, but simultaneously moves S at a constant speed of 3 m/s; D) falls in a straight line at a 45° angle to the vertical; E) falls in a straight line at a 30°angle to the vertical.

36 Still moving on your skateboard past your partners, you toss your car keys towards them, giving the keys a horizontal initial velocity of 3 m/s toward the south, relative to you. Relative to your partners, the keys: A) fall straight down; B) fall down, but simultaneously move N at a constant speed of 3 m/s; C) fall down, but simultaneously move S at a constant speed of 3 m/s; D) fall in a straight line at a 45° angle to the vertical; E) fall in a straight line at a 30°angle to the vertical.

Lab 3: Force and Motion

Equipment: push-and-pull spring scales, sensitive small force scales, hovercraft, masses, 2 strong magnets (1 large & 1 small), Rollerblade platform, tow rope and means of attaching a spring scale to platform, long stick for pushing platform, large rubber sling, protractor, steel ball and plastic encased magnet.

Free-Body Diagram A free-body diagram (FBD) is a drawing of a body, isolated from its surroundings, with all the forces acting on the body represented by force vectors, showing the magnitude and direction of the forces.

1 a) Use spring scales to observe the strength of pulls you exert with your hands. Forces are measured either in pounds or newtons (N). Exert forces of 1 N, 5 N, 10 N, 20 N, and 50 N. How does a force of 1 N compare to a force of 1 lb?

b) Use spring scales to observe the strength of pushes you exert with your hands. Exert forces of 1 N, 5 N, 10 N, 20 N, and 50 N.

Labs and Lab Concept Quizzes   **139**

2 Place the hovercraft on a level surface, table or floor, and turn it on. What does it take to move it? What does it take to keep it moving?

3 Turn the hovercraft off, still on the same surface. Now what does it take to move it? What does it take to keep it moving?
Compare this situation with **2** and try to explain any differences you observe. Be as precise as possible

4 What do you think would happen if you were to exert on opposite sides of the hovercraft two oppositely directed forces, so that the vector sum of the two forces is zero?

Try it. Was your prediction correct?

5 Suppose you were to apply a force to a hovercraft that is already moving, and the force you apply is perpendicular to the initial direction of motion. What do you think would happen?

Try it and describe the trajectory.

6 When a large magnet and a smaller magnet are placed close to each other, will one of the magnets have a greater force on it or will the forces on the two magnets be the same?
Observe the interaction between two such magnets. They are oriented so that they repel. Bring the magnets close together. How do the forces compare? Draw a FBD for each.

7 Use a force scale to measure the earth's pull on various masses: 0.1 kg (100 g), ½ kg (500 g), 1 kg, 2 kg. Use your measurements to formulate a force law for the earth's pull on a mass m. That is, formulate a simple rule whereby you could know the earth's pull on any mass. Check with your instructor or TA.

8 Force law for earth's pull: The force in N of earth's pull on a mass m equals the product of the mass and g, where g (an abbreviation for gravitational field) is a constant. The force of the earth's pull is called the object's weight, denoted by w.
$$w = mg$$

From your observations, what is the value of g? $g = $ _____ N/kg

Draw a FBD of a mass as its weight is being measured. Explain what body exerts each of the forces in your FBD.

Draw a FBD of the mass when it is not in contact with anything. What happens to the mass in this case?

9 Hold a 1 kg mass in the palm of your hand. What force or forces act on the object? Draw a FBD of the mass. Explain what body exerts each of the forces in your FBD.

10 Place the 1 kg mass on the table. Does the table exert a force on the mass?

How do you know?

Draw a FBD of the mass. Explain what body exerts each force in your FBD.

11 Push with your finger against your partner's arm. Did your finger exert a force on her arm?   Did her arm exert a force on you?
If so, how was the force related to the force you exerted, bigger, smaller, or the same?

12 Now extend your finger and have your partner push her arm into it. Did her arm exert a force on your finger?   In what way, if any, does this differ from your pushing your finger into her arm?

13 Go to PhET simulations on your computer. Under physics and motion, find the Gravity Force Lab Simulation, which shows the forces acting on two objects brought close together, the force of gravitational attraction, which is too small to be easily measured for objects of reasonable size. The mass of each object and the distance between the objects can be varied. How do the forces vary with distance?

What happens to the two forces if you increase the mass of either object?

What can you say in general about the relative magnitudes of the forces on the two objects?

What can you say about the directions of the forces on the two objects?

14 Open the PhET simulation Forces and Motion. Set the friction to ice (zero friction). Explore what happens when you have the man push on the box, exerting an applied force F_A. If the box is initially at rest and a constant force is applied to the box, what happens to the box? Observe carefully and describe the motion.

Start again with the box at rest and give the box a small push to the right and then stop pushing. Describe what happens to the box. You can play back the recording of the push and the motion to examine carefully.

Once the box is moving to the right, what could you do to get it to stop?

Can you summarize your observations with a simple statement about the force applied to the box and the motion of the box?

Now change the friction setting to wood, which means that there is significant friction. Describe in detail how this changes the relationship between the applied force and the motion of the block.

15 Seat one person on a Rollerblade platform. Another person pulls the platform with a tow rope. Attach two spring scales to opposite ends of the tow rope, attach one of the scales to the Rollerblade platform, and then pull on the other end as shown below. Begin very slowly and then move at a slow, constant speed. Use the scales to monitor the forces throughout the motion. Record your observations. How is force related to motion?
How do the forces on the platform and on the person pulling the platform compare?

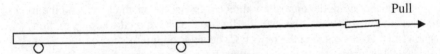

16 You need 3 people for this part. Two people pull on the ends of the sling and a third pulls in the middle in such a way that the sling is at rest. Use spring scales to measure the forces. Use chalk on the floor, a meter stick, and a protractor to measure angles. Draw a graphical construction of the vector addition of the 3 forces.

17 What happens to the trajectory of an object that is initially moving in the positive x direction when a constant force in the y direction is applied to the object?

Roll a steel ball along the length of the table, and after it is moving in a straight line, apply a constant force with a magnet. The force should be directed perpendicular to the motion. Was your prediction verified?

Lab 3 Concept Quiz: Force and Motion

1 How is force related to motion? Is a force required to start an object moving when that object is not initially moving? A) always; B) never; C) sometimes. If C, explain.

2 Is a force required to keep an object moving after it is initially moving? A) always; B) never; C) sometimes. If C, explain.

3 If a force is applied to an object that is initially moving and the force is directed perpendicular to the direction of motion, what effect will the force have on motion in the initial direction of motion? To be specific, if an object is initially moving in the x direction, what effect will a force on the object in the y direction have on the object's x component of velocity? A) increase v_x; B) decrease v_x; C) no effect on v_x.

4 If a force is applied to an object that is initially moving and the force is directed perpendicular to the direction of motion, what effect will the force have in the direction of the force, perpendicular to the initial velocity? A) no effect; B) causes motion in the direction of the force; C) causes motion in the direction opposite the force.

5 Which of the following can result in a force on an object: A) only i; B) only i and ii; C) only i, ii, iii; D) only i, ii, iii, iv; E) all: i, ii, iii, iv, v.

  i) pushing or pulling on the object by a human or other living being;

  ii) pushing or pulling by an inanimate thing;

  iii) a thing in contact with the object;

  iv) an object in contact with the earth or some other planet;

  v) an object near, but not in contact with the earth or some other planet.

6 If two equal magnitude, oppositely directed forces act on an object, and the forces are directed towards each other, what effect do they have on the body's motion? A) no effect; B) depends on the object; C) depends on the force.

7 Does a table exert a force on a stationary book on its surface? A) yes; B) no.

8 Does something that is in contact with an object always exert a force on it? A) yes; B) no. If no, give an example of a situation when contact results in no force.

9 Can something ever exert a force on an object without touching the object? A) yes; B) no. If yes, give an example.

10 If an object X exerts a force on an object Y, does Y exert a force on X? A) yes; B) no. If yes, can you make a general statement about how the magnitude of the force on X compares with the magnitude of the force on Y? If yes, can you make a general statement about how the direction of the force on X compares with the direction of the force on Y?

11 Which of the following are *not* possible effects of a force acting on an object? A) moves it; B) destroys it; C) spins it; D) deforms it; E) none of the above; all are possible effects.

12 What do the effects of a force on an object depend on in addition to the strength of the force? A) size or mass of the object; B) strength, rigidity of the object; C) other forces acting on the object; D) all of the above; E) none of the above.

13 The earth exerts a force on all objects on earth. That force is called the object's weight. Does an object's weight depend on its elevation? To be specific, would the weight of an object in this lab change significantly if you moved it to the third floor? A) yes; B) no.

14 Would the object's weight change if you moved it far out into space, millions of miles from earth? A) no; B) yes, it would increase a little; C) yes it would increase a lot; D) yes it would decrease a little; E) yes it would decrease to nearly zero.

Lab 4: Newton's Second Law

Equipment: hovercraft, one 2.5 N spring scale & screw to attach to hovercraft, 2m track, level, Pasco car, two 2.5 N spring scales, string, weights, brass spring, video camera and cable, tripod, LoggerPro software (adjust settings so that 3 points, not 7, are used for derivatives), computer.

1 Draw appropriate FBDs and predict what will happen when you give the hovercraft a very small push with your hand and let go. Describe in detail the motion you think the hovercraft will have both during the push and after it is no longer in contact with your hand.

Sketch your prediction of its velocity vs. time.

Now perform the experiment, while measuring the force and recording the motion. Attach a spring scale to the top of the hovercraft and apply a small force and then release, while recording the motion using the video camera; use LoggerPro to obtain a graph of v_x vs. t. Sketch v_x vs. t. Estimate the average force and how long it was applied.

Average force: _____

Duration of force: _____

How do your results compare with your predictions?

Discuss your results with the instructor.

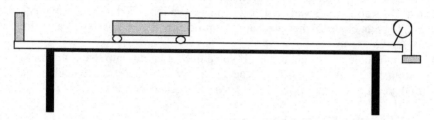

2 Consider the arrangement shown in the figure above: a low friction car on a horizontal 2-m track, with a spring scale attached to the top of the car, and string attached to the spring scale and to a hanging mass, as shown in the figure, so that there is tension in the string, a measurable force that is applied to the car. Initially the car is held in place. When it is released, the tension force pulls it along the track. After the mass m hits the floor, there is no longer tension in the string. There are two parts of the motion: 1) force pulling the car forward; 2) no horizontal force on car. Draw appropriate FBDs and predict what will happen when the car is released. Describe in detail the motion you think the car will have during each of the two parts.

Sketch your prediction of the car's velocity vs. time, showing the two parts. Then perform the experiment. Use a mass m that results in a force of 1 N on the front of the car while it is moving. (The force while held in place will be slightly more than 1 N.) Use the video camera and LoggerPro to record the motion, to obtain a graph of v_x vs. t, and to monitor the force applied to the car. Sketch v_x vs. t.

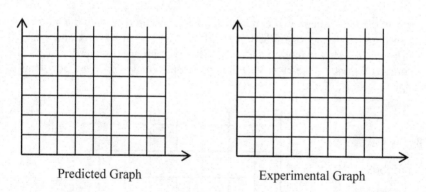

Predicted Graph   Experimental Graph

Analyze your results. How do your experimental results compare with your predictions?

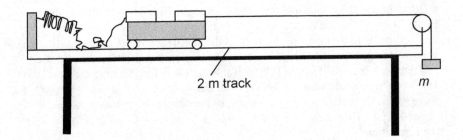

3 Consider the arrangement shown in the figure above: a low friction car on a horizontal 2 m track, with two spring scales attached to the top of the car and strings attached to the scales, so that when there is tension in a string, a measurable force is applied to the car. Initially the string attached to the rear of the car is slack so that there is no force applied there, but the string applied to the front of the car is taut because the string, after passing over a pulley, supports a mass m. After the mass m hits the floor, there is no longer tension in the front string. The length of the string attached to the back is such that it will be under tension after the mass hits the floor, but before the car reaches the end of the track. Initially the car is held in place. When the car is released, motion begins. There are 3 parts of the motion: 1) force pulling car forward; 2) no force on car; 3) force pulling car backward. Draw appropriate FBDs and predict what will happen when the car is released. Describe in detail the motion you think the car will have during each of the 3 parts.

Sketch your prediction of the car's velocity vs. time, showing the 3 parts.
Then perform the experiment. Use a mass m that results in a force of 1 N on the front of the car while it is moving. (The force while held in place will be slightly more than 1 N.) You will have to adjust the lengths of string so that you have the desired 3 types of motion. Use the video camera and LoggerPro to record the motion, to obtain a graph of v_x vs. t, and to monitor the forces applied to the car. Sketch v_x vs. t.

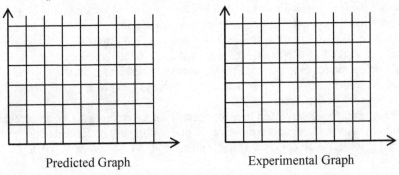

Predicted Graph   Experimental Graph

How do your experimental results compare with your predictions?

4 Go to the PhET website and open the Forces and Motion simulation. Select Force Graphs. Sketch an acceleration vs. time graph corresponding to whatever motion you want to impart to the crate in the simulation. Choose ice (no friction) to simplify the problem. Predict the graphs of force vs. time, velocity vs. time, and position vs. time. Sketch all of these graphs below before running the simulation. Once you have made your predictions, run the simulation, controlling the applied force to match your sketch. You can pause the simulation to make changes in the applied force. How do your predicted graphs compare with the graphs shown in the simulation?

This time make up a velocity vs. time graph for the motion you want to impart to the crate. Choose ice (no friction). Predict the graphs for acceleration, force, and position vs. time, based on your velocity graph. Then play the simulation, controlling the applied force to match your predicted graph. How do your graphs compare with the graphs shown in the simulation?

Lab 4 Concept Quiz: Newton's Second Law

1 What is the effect when a single force acts on an object that is initially not moving? A) object remains at rest; B) object starts moving.

2 What happens if a single force acts continuously on an object that is initially moving if the force is in the same direction as the initial velocity of the object? A) keeps moving at constant velocity; B) velocity keeps increasing for as long as the force acts; C) velocity increases for a while and then object moves at a higher constant velocity as force continues to act.

3 What happens if a single force acts continuously on a body that is initially moving if the force is directed opposite the initial velocity of the body? A) keeps moving at constant velocity; B) velocity decreases; if the force acts long enough, the object comes to rest and remains at rest even if the force continues to act; C) velocity decreases for as long as the force acts; if it acts long enough, the object moves in a direction opposite its original direction of motion.

4 What happens if two equal magnitude, oppositely directed forces act on an object that is initially at rest? A) object moves back and forth in the directions of the two forces; B) object moves randomly; C) object remains at rest and is crushed by the two forces; D) object remains at rest; whether it is crushed, depends on the object and the magnitude of the forces.

5 What happens if two equal magnitude, oppositely directed forces act on an object that is initially moving? A) object moves back and forth in the directions of the two forces; B) object soon comes to rest; C) object continues to move at constant velocity.

6 Two oppositely directed forces act on an object that is initially at rest. One force is directed to the right and a second, larger force is directed to the left. What happens? A) object moves back and forth in the directions of the two forces; B) object moves randomly; C) object remains at rest; D) object moves to the left; E) object moves to the right.

7 Two oppositely directed forces act on an object that is initially moving to the right. One force is directed to the left and a second, larger force is directed to the right. What happens? A) object slows down; B) object continues to move to the right at constant velocity; C) object continues to move to the right at an ever larger velocity; D) object accelerates to the right for a while, then continues to move at a larger constant velocity.

8 Two oppositely directed forces act on an object that is initially moving to the right. One force is directed to the right and a second, larger force is directed to the left. What happens? A) object slows down until it comes to rest and then stays at rest; B) object continues to move to the right at constant velocity; C) object slows down until it comes to rest and then begins to move to the left; D) object continues to move to the right at an ever larger velocity.

9 A force in the positive y direction acts on an object that is initially moving in the positive x direction. What happens? A) object immediately begins moving in the positive y direction; B) object moves at a constant velocity with components in both the x and y directions; C) object continues to move in the x direction; D) object begins to have a component of velocity in the y direction that continually increases, while having a constant velocity component in the x direction; E) object has velocity components in both the x and y directions & both continually increase.

10 Two equal magnitude forces, one in the positive x direction and one in the positive y direction, act on an object that is initially at rest. Object: A) begins to move along the x direction at increasing velocity; B) begins to move along the y direction at increasing velocity; C) begins to move in a straight line 45° above the x axis at increasing velocity; D) has no acceleration; E) moves along a crooked path, accelerating in both the x and y directions.

11 You are standing still on a diving board above a swimming pool. Draw a FBD. What forces act on you? A) only the force of gravity; B) the force of gravity directed down and a larger force of the earth pushing up; C) the force of gravity directed down and a larger force of the board pushing up; D) the force of gravity directed down and an equal force of the board pushing up; E) the force of gravity directed down and an equal force of the earth pushing up.

12 Now you step off of the diving board. Draw a FBD. At the instant you step off, what forces act on you? A) only the force of gravity; B) the force of gravity directed down and a larger force of the earth pushing up; C) the force of gravity directed down and a larger force of the board pushing up; D) the force of gravity directed down and an equal force of the board pushing up; E) the force of gravity directed down and an equal force of the earth pushing up.

13 You are on the ice in the middle of an ice hockey arena. A 100 g hockey puck is at rest in the middle of the ice. With your hand you give it a push, directed north, exerting a force on it of 5 newtons (N), about 1 pound, for 0.1 seconds. What happens to the puck during the 0.1 s interval you push it? Describe in detail. It might help to think of this by going through the motion in slow motion with your hand.

The force that keeps the puck going after the push (after contact with the hand) is:
A) the force of motion; B) the air; C) momentum; D) the push of the hand;
E) there is no force.

14 Which of the following graphs of v_y vs. time best describes the motion of the hockey puck in the last question?

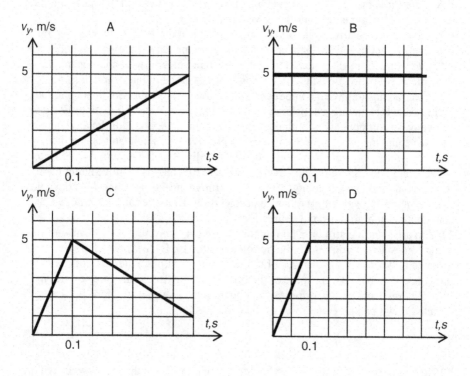

Lab 5: Projectiles

Equipment: video camera, computer, LoggerPro software, 2 steel balls with different masses, ramp and mount for table, mass balance, ruler, and meter stick.

1 Measure the mass of each of the two balls. $m_1 =$ _____ $m_2 =$ _____
Use Newton's second law to analyze and predict the motion of each of the balls when falling from rest. Draw appropriate FBDs for each of the masses, labeling all forces, identifying the source of each force, and indicating the value of each force. Then predict the acceleration of each mass and describe in words its predicted motion. Predict v_y vs. t for each.

2 With a vertical meter stick positioned nearby to facilitate measurement, use the video camera and LoggerPro to record the motion of the two balls as they are simultaneously dropped from the same height. Use LoggerPro to find the dot sequence of the motion and graphs of v_y vs. t and a_y vs. t for each of the balls. Sketch these results below. Discuss your results. Did they confirm your predictions?

3 Use Newton's second law to analyze and predict the motion of each of the balls when they are given a vertical upward initial velocity. Draw appropriate FBDs for each of the masses, labeling all forces, identifying the source of each force, and indicating the value of each force. Then predict the acceleration of each mass and describe in words its predicted motion. Predict v_y vs. t for each.

4 With a vertical meter stick positioned nearby to facilitate measurement, use the video camera and LoggerPro to record the motion of the two balls as they are simultaneously given the same upward velocity, starting from the same height. Use LoggerPro to find the dot sequence of the motion and graphs of v_y vs. t and a_y vs. t for each of the balls. Sketch these results below. Discuss your results. Did they confirm your predictions?

5 Use Newton's second law to analyze and predict the motion of each of the balls when they are given a horizontal initial velocity. Draw appropriate FBDs for each of the masses, labeling all forces, identifying the source of each force, and indicating the value of each force. Then predict the acceleration of each mass and describe in words its predicted motion. Predict v_x vs. t and v_y vs. t for each ball.

6 Adjust the ramp so that the end is at the edge of the table and is horizontal so that when a ball rolls down the track it will leave the track initially moving horizontally. With a vertical meter stick positioned nearby to facilitate measurement, use the video camera and LoggerPro to record the motion of the two balls as they are given the same horizontal initial velocity, starting from the same height. Use LoggerPro to find the dot sequence of the motion and graphs of v_x vs. t, v_y vs. t, a_x vs. t, and a_y vs. t for each of the balls. Sketch these results below. Discuss your results. Did they confirm your predictions?

7 Use Newton's second law to analyze and predict the motion of each of the balls when they are given an initial velocity directed above the horizontal. Draw appropriate FBDs for each of the masses, labeling all forces, identifying the source of each force, and indicating the value of each force. Then predict the acceleration of each mass and describe in words its predicted motion. Predict v_x vs. t and v_y vs. t for each ball.

8 Adjust the track so that the bottom end of the track is angled upward about 20° to 30°. With a vertical meter stick positioned nearby to facilitate measurement, use the video camera and LoggerPro to record the motion of the two balls as they are given the same initial velocity directed above the horizontal, starting from the same height. Use LoggerPro to find the dot sequence of the motion and graphs of v_x vs. t, v_y vs. t, a_x vs. t, and a_y vs. t for each of the balls. Sketch these results below. Discuss your results. Did they confirm your predictions?

Lab 5 Concept Quiz: Projectiles

A projectile is an object that is projected or thrown – something that is given an initial velocity so that it then moves through the air. In the questions that follow assume that the projectile is any object of mass m, a ball for example, that starts at some height above the floor, is either dropped or thrown, and moves through the air with no significant air resistance until it finally hits the floor. The x axis is in the horizontal direction and the y axis is in the upward direction.

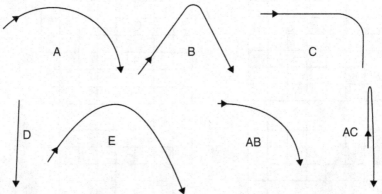

1 An object is released from rest. Which of the sketches above best describes the object's trajectory?

2 An object is thrown vertically upward, i.e., it is given an initial velocity in the upward direction. Which of the sketches above best describes the object's trajectory?

3 An object is thrown horizontally, i.e., it is given an initial velocity in the horizontal direction. Which of the sketches above best describes the object's trajectory?

4 An object is thrown upward and outward, i.e., it is given an initial velocity directed above the horizontal. Which of the sketches above best describes the object's trajectory?

A	B	C	D
4 ○	0 ○	0 ○	0 ○
3 ○	1 ○		1 ○
2 ○	2 ○	1 ○	
			2 ○
1 ○	3 ○	2 ○	
		3 ○	3 ○
0 ○	4 ○		
		4 ○	4 ○

5 Which of the sequences of dots in the preceding figure, which show positions at equal time intervals, best describes the motion of an object released from rest?
6 Which of the dot sequences in the preceding figure best describes the motion of an object thrown upward?
7 Which of the dot sequences below best describes the motion of an object thrown horizontally?

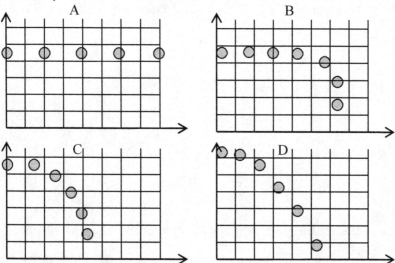

8 Which of the dot sequences below best describes the motion of an object thrown with an initial velocity directed above the horizontal?

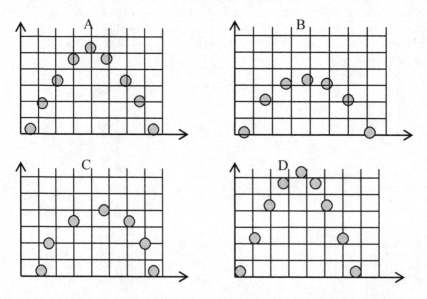

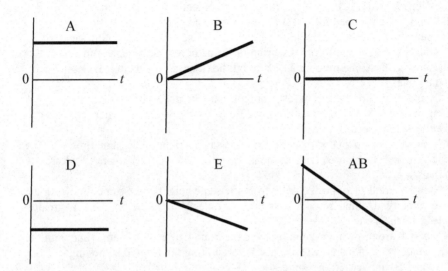

9 An object falls from rest. Which of the graphs above best represents velocity v_x vs. t?

10 An object falls from rest. Which of the graphs above best represents velocity v_y vs. t?

11 An object falls from rest. Which of the graphs above best represents acceleration a_x vs. t?

12 An object falls from rest. Which of the graphs above best represents acceleration a_y vs. t?

13 An object is thrown horizontally. Which of the graphs above best represents v_x vs. t?

14 An object is thrown horizontally. Which of the graphs above best represents v_y vs. t?

15 An object is thrown horizontally. Which of the graphs above best represents a_x vs. t?

16 An object is thrown horizontally. Which of the graphs above best represents a_y vs. t?

17 An object is thrown with an initial velocity directed above the horizontal. Which of the graphs above best represents v_x vs. t?

18 An object is thrown with an initial velocity directed above the horizontal. Which of the graphs above best represents v_y vs. t?

19 An object is thrown with an initial velocity directed above the horizontal. Which of the graphs above best represents a_x vs. t?

20 An object is thrown with an initial velocity directed above the horizontal. Which of the graphs above best represents a_y vs. t?

Questions 21 – 31: Object X has mass m and object Y has mass $2m$.

21 X and Y are dropped from rest from the same height at the same time. Compared with X, the acceleration of Y is: A) greater; B) less; C) the same.

22 X and Y are dropped from rest from the same height at the same time. Compared with X, the time it takes Y to hit the floor is: A) greater; B) less; C) the same.

23 X and Y are dropped from rest from the same height at the same time. Compared with X, just before hitting the floor Y is moving: A) faster; B) slower; C) at the same speed.

24 X and Y are thrown horizontally from the same height at the same time with the same velocity. Compared with X, the acceleration of Y is: A) greater; B) less; C) the same.

25 X and Y are thrown horizontally from the same height at the same time with the same velocity. Compared with X, the time it takes Y to hit the floor is: A) greater; B) less; C) the same.

26 X and Y are thrown horizontally from the same height at the same time with the same velocity. Compared with X, just before hitting the floor Y is moving: A) faster; B) slower; C) the same.

27 X and Y are thrown horizontally from the same height at the same time with the same velocity. Compared with X, the horizontal distance Y travels before hitting the floor is: A) greater; B) less; C) the same.

28 X and Y are thrown from the same height at the same time with the same velocity, directed above the horizontal. Compared with X, the acceleration of Y is: A) greater; B) less; C) the same.

29 X and Y are thrown from the same height at the same time with the same velocity, directed above the horizontal. Compared with X, the time it takes Y to hit the floor is: A) greater; B) less; C) the same.

30 X and Y are thrown from the same height at the same time with the same velocity, directed above the horizontal. Compared with X, just before hitting the floor Y is moving: A) faster; B) slower; C) the same.

31 X and Y are thrown from the same height at the same time with the same velocity, directed above the horizontal. Compared with X, the horizontal distance Y travels before hitting the floor is: A) greater; B) less; C) the same.

Questions 32 – 35: object Y has an initial velocity that is twice that of object X; both are given an initial velocity in the same direction above the horizontal and both start from the same elevation at the same time.

32 Compared to X, the acceleration of Y is: A) greater; B) less; C) the same.

33 Compared to X, the maximum height of Y is: A) greater; B) less; C) the same.

34 Compared to X, the time it takes Y to hit the floor is: A) greater; B) less; C) the same.

35 Compared to X, the horizontal distance Y travels before hitting the floor is: A) greater; B) less; C) the same.

Use the FBDs below to answer **questions 36 – 39**. **P** means a push & **w** means weight, the force of gravity.

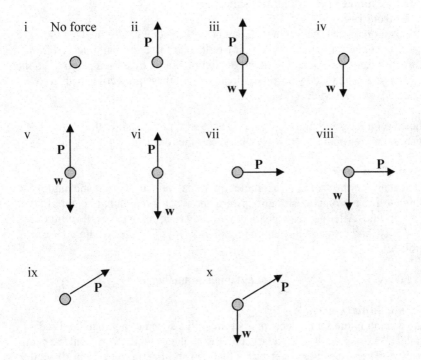

36 An object is released from rest. After its release, the FBD of the object is:
A) i throughout its motion; B) iv throughout its motion; C) iv at the beginning of its motion and ii just before it hits; D) i initially, then iv.

37 A ball is thrown vertically upward. The FBD of the ball after it leaves the hand is: A) ii while it is going up and iv while it is going down; B) ii while going up, i at the highest point, and iv while going down; C) i throughout its motion; D) iv throughout its motion; E) iv at the beginning of its motion and ii just before it hits.

38 A ball is thrown horizontally. The FBD of the ball after it leaves the hand is: A) vii throughout its motion; B) viii throughout its motion; C) vii initially and then iv; D) iv throughout its motion; E) iv at the beginning of its motion and ii just before it hits.

39 A ball is thrown with an intial velocity directed above the horizontal. The FBD of the ball after it leaves the hand is: A) ix while it is going up and x while it is going down; B) ix while going up, i at the highest point, and x while going down; C) i throughout its motion; D) iv throughout its motion; E) iv at the beginning of its motion and ii just before it hits.

Lab 6: Projectile Simulations

Equipment: computer, PhET website: physics, motion, projectile motion
Use the website to simulate the motion of falling bodies and projectiles.

I No Air Resistance (Turn off air resistance.)
1 Free fall from rest
Raise the cannon so that the origin is 20 m above the ground. Use the tape measure to measure the height. Predict how long it would take for a golf ball released from this height to reach the ground. Predict how long it would take for a piano to do the same. After you have made your predictions, release these projectiles with zero initial velocity and observe.

Predicted time for golf ball _____   Observed time for golf ball _____
Predicted time for piano _____   Observed time for piano _____

2 Vertical initial velocity
With the cannon in the same place, predict the initial velocity (speed and angle) for the cannon to fire a projectile straight upward and reach a maximum height of 50 m above the ground. After you have made your prediction, fire a projectile with this initial velocity, and measure its maximum height. Does it matter what you use as your projectile?

Initial velocity: _____ _____   Measured maximum height _____

3 Horizontal initial velocity
With the cannon in the same place, predict the initial speed of a projectile, fired horizontally, in order for it to hit a target that is on the ground, 20 m from the center of the tower. Then place the target at this location and fire a projectile horizontally with this initial speed.

Predicted initial speed _____   Did you hit the target? _____

4 Non-horizontal initial velocity
With the cannon in the same place, predict the initial velocity required to hit a target at a point P at the same elevation as the cannon and 20 m away from the origin. Find the initial velocity, speed and angle, for which the speed is least. Also predict how far the projectile will move in the horizontal direction before hitting the ground. Place the target at point P and fire the cannon, giving the projectile your computed initial velocity.

Predicted initial velocity _____ _____
Predicted total horizontal distance traveled _____
Did you hit the target? _____
Observed total horizontal distance traveled _____

5 Finding an initial speed and angle to hit a target
Now place the cannon as close to the ground as possible and place the target at $x = 20$ m, $y = 20$ m. (Use the tape measure.) Calculate an initial velocity (speed and angle) of a projectile that will hit the target. Hint: Choose a reasonable angle first and then calculate the initial speed. Repeat this with a different value of the angle.

First predicted initial velocity _____ _____ Did you hit the target with this velocity? _____
Second predicted initial velocity _____ _____ Did you hit the target with this velocity? _____

6 Finding a target position with a given initial velocity
With the cannon in the same place, and with a target at $y = 10$ m, predict two values of the x coordinate of the target, such that a projectile with an initial velocity of 20 m/s, directed 60° above the horizontal, will hit the target. Then place the target at each of these locations and see if the projectile hits the target at these locations, with the given initial velocity.

Target position 1, x_1 = _____ Did the projectile hit the target at this position?
Target position 2, x_2 = _____ Did the projectile hit the target at this position?

7 Your own projectile problem
Create your own projectile problem, solve it, and test your prediction using the simulation.

II Air Resistance (Turn on air resistance.)

Free Fall from Rest. (Set the initial velocity to zero.)

1 Predicted small effect of air resistance
Consider various combinations of projectile mass, diameter, and drag coefficient. Choose a set of values that you think will give a negligible effect of air resistance. Then try these values. What can you see that shows whether air resistance is large or small? What can you do to show quantitatively whether air resistance is negligible? Was air resistance small? If it was small, was it negligible?

2 Predicted large effect of air resistance
Consider various combinations of projectile mass, diameter, and drag coefficient. Choose a set of values that you think will give a large effect of air resistance. Then try these values. What can you see that shows whether air resistance is large or small? What can you do to show quantitatively whether air resistance is significant or negligible? Was air resistance large? If it was not large, was it significant?

Lab 6 Concept Quiz: Projectiles, part 2

A ski jumper wants to have a longer jump, so he puts on a heavy back pack, reasoning that an increase in his mass will give him more velocity and make him travel farther. Critique his reasoning, assuming no air resistance and no friction on the track.

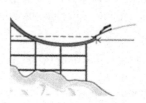

1 As a result of the back pack, the distance he travels before hitting the ground will be:
A) greater; B) less; C) the same.
2 The time he is in the air will be: A) greater; B) less; C) the same.
3 His acceleration while in the air will be: A) 9.8 m/s^2; B) less than 9.8 m/s^2; C) greater than 9.8 m/s^2.

An archer wants to be able to shoot an arrow so that it hits the ground as far as possible from the point where he shoots it. The ground is level. The archer reasons that the arrow should be aimed almost horizontal, so that its velocity component along the horizontal is as great as possible, and therefore will travel as far as possible in the horizontal direction. Critique her reasoning.
4 Is she right? A) yes; B) no, because this won't make the horizontal component of velocity bigger; C) no, even though this will make the horizontal component of velocity bigger.
5 A second archer argues that in order to travel as far as possible before hitting the ground, the arrow should be shot almost vertically upward, so that the arrow is in the air as long as possible and therefore will travel farther. Is he right? A) yes; B) no, because this won't keep the arrow in the air longer; C) no, even though this will keep the arrow in the air longer.
6 For maximum distance, the angle that the arrow makes above the horizontal as it is shot should be close to: A) 10°; B) 30°; C) 45°; D) 60°; E) 80°.
7 Archers are positioned at the top of a very high castle wall, surrounded by a flat plain. In order for their arrows to strike the ground as far as possible from the castle, they should aim their arrows above the horizontal at: A) an angle < 45°; B) an angle = 45°; C) an angle > 45°.
8 At a certain point in a projectile's trajectory, the tangent to the trajectory is at an angle of 60° with the horizontal. If the projectile's horizontal component of velocity is 10 m/s, its speed is: A) 10 m/s; B) 20 m/s; C) 30 m/s; D) 5 m/s; E) - 10 m/s.
9 A passenger in a convertible traveling along a horizontal road at 3 m/s throws a ball vertically upward and catches it. Air resistance is negligible. Relative to the passenger, the ball's horizontal component of velocity while it is in the air is:
A) 3 m/s; B) - 3 m/s; C) 0. Sketch the trajectory seen by the passenger.
10 A passenger in a convertible traveling along a horizontal road at 3 m/s throws a ball vertically upward and catches it. Air resistance is negligible. Relative to a person standing on the sidewalk, the ball's horizontal component of velocity is:
A) 3 m/s; B) - 3 m/s; C) 0. Sketch the trajectory seen by the standing person.

Lab 7: Friction
Equipment: 2 m track, wooden block and plastic block with strings attached, spring scales, two 1 kg masses, level, lab jack to raise one end of track.
I Experiments

1 Level the track. Place the wooden block on the 2 m track, and place the 1 kg mass on top of the block. Using a spring scale to measure the force, pull the block horizontally to the right with a force of 2 N. What happens?
Choose a free body & draw a free body diagram. Apply Newton's secnd law.

FBD

Name and evaluate all the forces on the free body. Are the forces related in any way? How do you know?
What is the magnitude of the friction force? _____ How do you know?

2 Repeat the experiment, but this time pull with a force just big enough to move the block at a slow, steady rate. How big is this force? ____ Draw a new free-body diagram. Apply Newton's second law.

FBD

Name and evaluate all the forces on the free body. Are the forces related?
What is the magnitude of the friction force? _____ How do you know?

3 Predict what would happen if you were to repeat part **2**, but with 2 kg placed on the block instead of 1 kg.. Draw a free body diagram, apply N's 2nd law & force laws. Use your analysis to predict how big a force you must apply now to get the block to move. Once you have made your prediction, do the experiment.

FBD

Predicted force required to move block: _____
Observed force required to move block: _____

4 Predict what would happen if you were to repeat part **2**, with the 1 kg mass placed on the block, but with a force twice as great as the one you applied in part **2**. Draw a free body diagram, apply N's 2nd law & force laws. Use your analysis to predict what would happen. Once you have made your prediction, do the experiment.

FBD

Prediction: _____

Observation: _____

5 Predict what would happen if you were to repeat part **1**, replacing the wooden block by a plastic block. Do you think the new block will still remain at rest when pulled by a 2 N force? Explain the reason for your prediction.

Do the experiment. Was your prediction verified?

6 Write a paragraph explaining what the experiments you have done in this lab show about the nature of the friction force. Before going on discuss your results with the instructor.

II Simulations In this part you will repeat what you did with real objects in part I, but this time with simulated objects, 100 times more massive, and with automatic graphing of v_x vs t. On your computer find the PhET web site. Find the Motion category and open the simulation: Forces in 1 Dimension. Choose the crate as the object you will push on. Click on "more controls" and adjust the mass of the crate to 100 kg, roughly 100 times the mass of the real object you pushed on in the previous parts. Set the coefficients of static and kinetic friction equal to 0.3, and set the gravitational field to 10 N/kg. Display graphs of force vs. time, velocity vs. time, and position vs. time. Note: The simulation plays in slow motion. The time recorded on the graphs is roughly half as long as the actual elapsed time as the simulation plays. The crate is initially at $x = -7$ m.
1 Analyze the crate and predict the applied force required to move it. Draw a series of four FBDs with increasing applied forces, with the last applied force just able to move the crate.

Once you have made your prediction and drawn your FBDs, run the simulation with the applied forces in each of the FBDs. What do you observe?

2 Repeat **1** with the coefficient of friction doubled. Physically, what have you changed?

3 Apply a steady force to the right sufficient to move the crate at a constant velocity of 1 m/s. This may require a few tries. What is the relationship between the applied force and the friction force? Is this consistent with Newton's second law? Explain. Draw a FBD.

4 Predict the applied force that would cause the crate to increase its velocity 1 m/s each second for 10 s. Sketch the graphs of v_x vs. t and x vs. t. Draw a FBD and use the data from the previous part.

After you have made your prediction, run the simulation. Were you right?

5 Suppose you want to move the crate 16 m to the right and have it at rest at that point without overshooting it. Analyze forces and create a plan for the variable force you will apply at each point along the way. Make your prediction. Predict graphs of force vs time, velocity vs time, and position vs time, Then run the simulation. Pause the simulation whenever you adjust the value of the applied force, adjusting the force according to your plan. Be careful that your plan has the crate moving slowly enough that you will be able to pause where you want to. Did your plan work? If not, try again.
Discuss your results with the instructor.

Lab 7 Concept Quiz: Friction

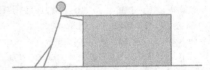

1 A heavy table is at rest on a horizontal floor. Nothing touches the table except the floor. Is there a friction force exerted on the table by the floor? A) yes, directed to the left; B) yes, directed to the right; C) no. Draw a FBD of the table. Show all forces and sources of those forces.

2 You exert a horizontal force to the right on the table, but it is not enough to move it. Is there a friction force exerted on the table by the floor? A) yes, directed to the right, less than the push; B) yes, directed to the left, less than the push; C) yes, directed to the right, equal to the push; D) yes, directed to the left, equal to the push; E) no. Draw a FBD of the table. Show all forces and sources of those forces.

3 You now push harder on the table and it begins to move to the right. As its motion begins, is there a friction force exerted on the table by the floor? A) yes, directed to the right, less than the push; B) yes, directed to the left, less than the push; C) yes, directed to the right, equal to the push; D) yes, directed to the left, equal to the push; E) no. Draw a FBD of the table. Show all forces and sources of those forces.

4 You continue pushing the table, and it moves to the right at constant velocity. Is there a friction force exerted on the table by the floor? A) yes, directed to the right, less than the push; B) yes, directed to the left, less than the push; C) yes, directed to the right, equal to the push; D) yes, directed to the left, equal to the push; E) no. Draw a FBD of the table. Show all forces and sources of those forces.

5 You stop pushing the table. At the instant you stop pushing, the table is moving to the right. What happens next? A) instantly stops because there is no longer any horizontal force acting on it; B) instantly stops because there is a friction force acting on it; C) decelerates and quickly comes to rest because there is a friction force acting on it. Draw a FBD of the table. Show all forces and sources of those forces.

6 A heavy box is now placed on the table. Nothing else touches the table except the floor. Is there a friction force exerted on the table by the box? A) yes, directed to the right; B) yes, directed to the left; C) no. Draw a FBD of the table and box together. Show all forces and sources of those forces.

7 A heavy box is on the table. Nothing else touches the table except the floor. Is there a normal force exerted on the table by the box? A) yes, directed down, equal to the weight of the table; B) yes, directed down, equal to the weight of the box; C) yes, directed up, equal to the weight of the box; D) yes, directed up, equal to the weight of the table; E) no. Draw separate FBDs of the table and the box. Show all forces and sources of those forces.

8 A heavy box is on the table. You exert a horizontal force to the right on the table, but it is not enough to move it. Is there a friction force exerted on the table by the floor? A) yes, directed to the right, less than the push; B) yes, directed to the left, less than the push; C) yes, directed to the right, equal to the push; D) yes, directed to the left, equal to the push; E) no. Draw a FBD of the table and box together. Show all forces and sources of those forces.

9 A heavy box is on the table. You now push harder on the table and it begins to move to the right. As its motion begins, how does the friction force compare with the friction force in **3**, when there was no box on the table? A) same friction force as in **3**; B) smaller friction force than in **3**; C) larger friction force than in **3**. Draw a FBD of the table and box together. Also draw separate FBDs of the table and box. Show all forces and sources of those forces.

10 A heavy box is on the table. You continue pushing the table, and it moves to the right at constant velocity. How does the friction force compare with the friction force in **4**, when there was no box on the table? A) same friction force as in **4**; B) smaller friction force than in **4**; C) larger friction force than in **4**. Draw a FBD of the table and box together. Also draw separate FBDs of the table and box. Show all forces and sources of those forces.

11 A heavy table is at rest on a ramp that is inclined at some angle with the horizontal. Nothing is in contact with the table except for the surface of the ramp. Is there a friction force exerted on the table by the ramp? A) yes; B) no. Draw a FBD of the table. Show all forces and identify the sources of those forces.

Lab 8: Circular Motion

Equipment: video camera, computer, LoggerPro software, 2 rulers, steel ball and turntable, stand and pendulum support, string and pendulum mass of 150 g, meter stick, large protractor, marble and foam track. One per room: loop-the-loop track and appropriate ball.

I Acceleration During Circular Motion at Constant Speed

An object is moving at a constant speed of 30 cm/s along a circular path of radius 6 cm. You will find the object's acceleration at the instant it is at point P by graphical construction. (Figure should be enlarged to radius of 6 cm.)

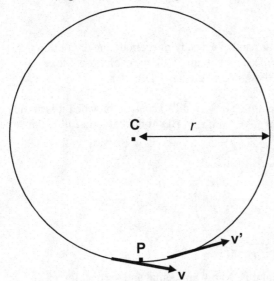

Use this figure to construct the change in velocity during a short time interval while the object is near point P in its circular trajectory. Velocities **v** and **v'** are shown just before and just after the object is at point P. Begin by drawing the tail of vector **v'** at point P and constructing the change in the velocity $\Delta \mathbf{v} = \mathbf{v'} - \mathbf{v}$ during the time interval, which is 0.1 s. In what direction does $\Delta \mathbf{v}$ (and therefore also acceleration) point?

Measure the magnitude of $\Delta \mathbf{v}$, using the scale for the velocity vectors (1cm = 10 cm/s) and compute the magnitude of the object's acceleration.

Measured $a =$ _____

Compute v^2/r. $v^2/r =$ _____

In uniform circular motion, acceleration is always directed toward the center of the circle and has magnitude v^2/r.

II Ball on a Turntable

1 Circular motion will be imparted to a steel ball by placing it at the outer edge of a turntable, where it is held in place by a magnet embedded in the turntable. As the turntable turns, the ball moves along a circular path. Measure the mass of the ball.

$m =$ _____

Rotate the turntable with your hand at a steady rate, about one revolution per second. Use a stopwatch to measure the time for one revolution. Make other measurements to determine the speed of the ball. Time for one revolution = _____
Speed of ball = _____

Predict the direction and magnitude of the ball's acceleration when it is at the point in its trajectory shown below. Also draw a FBD of the ball, indicating the value of all forces and the source of those forces.

P.

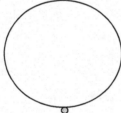

Draw a FBD viewed from point P. Show horizontal and vertical forces.

2 When you stop the turntable with your hand, the ball falls off. The motion will be shot with a video camera above the turntable and the motion analyzed with LoggerPro. But first predict the motion of the ball as it stays attached through ¾ of a complete circle and then falls off. Draw images of the ball at equal time intervals. Also show in each of the images in your sketch, horizontal forces, if any, acting on the ball. An image of the turntable is provided for reference.

1st image of ball

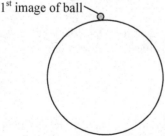

Do the LoggerPro images confirm your prediction?

III The Simple Pendulum

A small mass m suspended by a string is called a simple pendulum. If the mass is moved to the side so that the string is not vertical and then released, the pendulum swings back and forth.

1 What is the shape of the trajectory of the mass m? Sketch the trajectory in the figure. What is the direction of the mass's velocity when it is at the lowest point?

2 Draw FBDs of the mass at several points in its trajectory. Indicate all forces and the sources of those forces.

3 By analyzing the forces acting on the mass, find an expression for the ratio of the tension in the string to the weight of the mass when the angle the sting makes with the vertical has its greatest value, θ_{max}. Use Newton's second law in component form, choosing coordinates wisely.

At $\theta = \theta_{max}$, $T/(mg) = $ _____   If $\theta_{max} = 45°$, $T/(mg) = $ _____

4 Draw a FBD of the mass when it is moving at an instant when the string is vertical. Identify the forces. Are the forces equal? Apply Newton's second law and what you know about this kind of motion.

5 If the maximum angle that the pendulum string makes with the vertical is 45°, then the speed of the mass is $(0.6\, gl)^{1/2}$ when the string is vertical, where l is the length of the string. (We will know how to prove this after we learn about energy.)

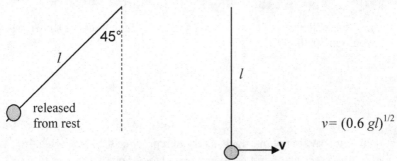

If the mass $m = 150$ g, what is the tension in the string when the string is vertical and when it is at 45°?
Calculated values: $T_{vert} = $ _____ $T_{45°} = $ _____

6 Form a simple pendulum with a string, a 2.5 N spring scale to measure string tension, and a 150 g mass. Release the pendulum at an initial angle of 45°. Observe the spring scale at the bottom and at the endpoints of its arc.
Measured values: T_{vert} = _____ $T_{45°}$ = _____
Are your predictions for tension verified?

IV PhET Pendulum Lab Open the PhET pendulum simulation. Show velocity and acceleration vectors. Play the simulation, first in real time, then at 1/16-th time, so that you can observe how velocity and acceleration are changing throughout the motion. Sketch FBDs of the mass at three different positions: bottom, highest point, and intermediate point. Explain how forces in your FBD account for the observed acceleration vector and how velocity and acceleration are related at each of the three points.

Bottom   High Point   Intermediate Point
At which of the three positions is acceleration solely a rate of change in the direction of the velocity?
At which of the three positions is acceleration solely a rate of change in the magnitude of the velocity?

V Loop-the-Loop
1 Play with the loop-the-loop at the front of the room and determine experimentally the initial height of the ball, expressed in terms of the loops radius R, in order for it to complete the loop. What forces act on the ball?

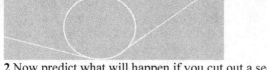

2 Now predict what will happen if you cut out a section of the track, as shown below. What would the path of the ball be?

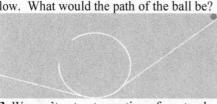

3 We can't cut out a section of our track, so try to reproduce the same effect by using a marble and a section of curved foam track. Is your prediction verified?
VI Cup of Water Moving in a Vertical Circle
(Carefully) swing the cup of water on a platform in a vertical circle. Why doesn't the water fall out when the cup is upside down?

Lab 8 Concept Quiz: Circular Motion

1 You are driving around a curve and encounter a patch of ice. Assuming there is no friction at all, which of these paths will your car follow? Explain, drawing FBDs of the car on the non-icy and on the icy parts of the road.

2 When your car is in the initial position shown in the figure, turning to the left along the curve of the road, the acceleration of you and your car is: A) zero; B) directed to the left; C) directed to the right.

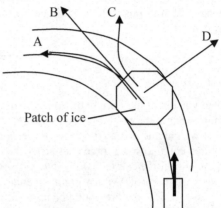

3 When your car is in the initial position shown in the figure, turning to the left along the curve of the road, the only horizontal force on you and your car is: A) zero – there is no horizontal force; B) the friction force of the road on the tires to the left; C) the friction force of the road on the tires to the right; D) the push of the road on the tires in the forward direction; E) the friction force of the road on the tires and a reaction force that cancels it.

4 When your car is on the icy section of road, the only horizontal force on you and your car is: A) zero – there is no horizontal force; B) the friction force of the road on the tires to the left; C) the friction force of the road on the tires to the right; D) the push of the road on the tires in the forward direction; E) the friction force of the road on the tires and a reaction force that cancels it.

5 You are on a moving merry-go-round. Instead of riding normally, you are hanging onto a strap from the top, as shown. You then let go. Describe your motion after you let go.

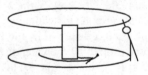

Viewed from a tower directly above the point where you let go, which of the following paths is closest to the path your body follows?

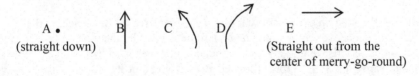

A. (straight down)   B   C   D   E (Straight out from the center of merry-go-round)

6 Viewed by a person standing on the ground beside the point where you let go, which of the following paths is closest to the path your body follows? Explain, drawing FBDs of your body while in contact with the merry-go-round and after you break contact.

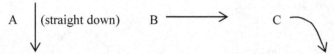

7 When you are in the initial position shown in the figure, still holding onto the merry-go-round, the only forces on you are: A) zero – there are no forces on you; B) your weight and an equal upward force exerted by the merry-go-round; C) your weight, an equal upward force exerted by the merry-go-round, and a horizontal component of force exerted by the merry-go-round, directed toward the center of the merry-go round; D) your weight, an equal upward upward force exerted by the merry-go-round, and a horizontal component of force exerted by the merry-go-round, in the direction you are moving; E) only a force directed away from the center of the merry-go-round.

8 After you let go of the merry-go-round, but while you are still in the air, the only forces on you are: A) zero – there are no forces on you; B) your weight and an equal upward force exerted by the merry-go-round; C) your weight, an equal upward force exerted by the merry-go-round, and a horizontal component of force exerted by the merry-go-round, directed toward the center of the merry-go-round; D) your weight; E) only a force directed away from the center of the merry-go-round.

9 Tarzan leaps from a tree and swings from a vine, which at some point breaks. Just before the vine breaks, the only forces on Tarzan are: A) his weight and an equal upward force exerted by the vine; B) a force exerted by the vine, directed toward point P; C) his weight; D) his weight and a force exerted by the vine, directed toward point P; E) a force directed away from point P.

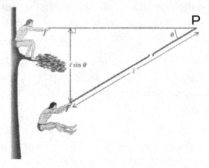

10 Just after the vine breaks, the only forces on Tarzan are: A) his weight and an equal upward force exerted by the vine; B) a force exerted by the vine, directed toward point P; C) his weight; D) his weight and a force exerted by the vine, directed toward point P; E) a force directed away from point P.

Draw FBDs of Tarzan before and after the vine breaks.

11 The girl on the swing is at the highest point in her motion at the instant shown and is instantaneously at rest. The forces acting on her at that instant are: A) her weight and an equal upward force exerted by the chain; B) her weight and a pull exerted by the chain, upward and to the left, parallel to the chain; C) her weight; D) no force acts on her.

12 The resultant force on the girl at the instant shown is: A) zero; B) directed vertically downward; C) directed upward and to the left, parallel to the chain; D) downward and to the left, perpendicular to the chain.

13 The acceleration of the girl at the instant shown is: A) zero; B) directed vertically downward; C) directed upward and to the left, parallel to the chain; D) downward and to the left, perpendicular to the chain.

14 The tension in the chains at the instant shown is: A) zero; B) equal to the weight; C) equal to the component of the weight parallel to the chain. Draw a complete FBD of the girl.

Lab 9: Newton's Second Law, Rollerblade Platform

Equipment: Rollerblade platform (plywood, mounted on very low friction Rollerblade wheels) & long tow rope, 50 N spring scale, metronome, washable markers, sugar packets, 30 m tape measure.

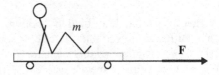

1 Predict qualitatively the motion that would result from either you or your lab partner being pulled, while sitting on the platform, with a constant horizontal force of about 20 N. (The value of the force you should use depends on your weight.)

Qualitative run Find a hallway or sidewalk that is flat, level, and clear of traffic over a distance of 15 to 20 m. One student sits on the platform and the other pulls the platform through the hall in a straight line with a constant force. The tow rope should be long enough that it is nearly horizontal when held at a convenient height. A spring scale attached to the end of the rope reads the force applied to the rope. The platform must be pulled in such a way that the reading of the spring scale is as constant as possible. If the force is not constant, you are not moving correctly. Think about how you have to move in order to keep the force constant. Switch roles, so that each member of the group pulls the cart. How did you move so as to keep the force constant? (It is not possible to keep the force perfectly constant, but if you move the right way you should be able to observe a force that fluctuates a few newtons above or below the desired value, so that the average value is quite close to that value.)

2 Quantitative analysis First analyze the motion theoretically. Then measure the motion and compare the results of your measurement with your theoretical prediction. What would a complete prediction of the motion entail?   What must you be able to predict?
You should be able to compare your prediction with measurements you will make. Work through the analysis and develop a general equation that completely specifies the motion. Later you will graph that equation, using specific values of tension, mass, and other variables.

Sketch of platform pull	**FBD**

Law(s), equations used in analysis:

General equation describing the motion, x vs. t: _____
In addition to the variables describing motion (x and t), your general equation should contain the variables: tension T, friction f, mass m, and initial velocity v_{x0}.
   You will apply a specific value of tension when you perform the experiment, but first you should determine f and m. Explain how you will determine these quantities, then do so and record their values. Draw FBDs as needed.

Who will ride? _____
How to determine m: Value of m:_____

Please note: in this experiment the rolling friction is velocity dependent! For best quantitative results, analyze the motion only after the cart is moving for 5 s, after which the friction force is nearly constant. How will you measure friction for this part of the motion?

Value of f:_____   Value of T: _____
Substitute specific values into your general equation, and obtain an equation for x as a function of t that describes the specific motion you will experience in the experiment. It should also involve initial velocity v_{x0} at the beginning of the measured motion, after 5 s. Equation describing your motion in this experiment:

_____

3 Experiment Perform the experiment described in **2**. Measure time with the metronome. Drop a sugar packet or make a mark every second, beginning 5 s after the motion begins. Take the first packet or mark as your origin and measure the distance x to each mark (not the distance between marks).

Helpful hints: a) The person who is riding should both hold the metronome timer and mark the position at 1 s intrvals. Why?
b) The rider should be positioned with heels extended over the front edge of the cart, to serve as brakes. The rider should hold the timer in one hand and the marker or sugar packets in the other hand, which should be held beside a fixed point on the cart, just above the level of the floor, so that marks can be made easily and quickly.
c) The run should last no more than 8 s after the 5 s to avoid dangerous speeds.
d) Choose a tension that is big enough to reduce experimental error, but not so big that the cart will be out of control.
e) For best results do the experiment two or three times to see if your results are consistent. You don't need to measure separately three distances to average the results. Measuring to the midpoint of the three marks accomplishes the same thing more quickly.

Perform the experiment, measure, and record your data in the data table.
Use your first two data points to estimate v_{x0}: v_{x0} _____

t, s	x, m
0	0.00
1	
2	
3	
4	
5	
6	
7	
8	

4 Analysis Compare your theoretical prediction of the motion with your experimental data describing the motion. Do this by plotting the equation found in **2** along with the data found in **3** on a single graph, using Data Studio. Print your graph if possible. If not, show the results in a graph below. Using Data Studio: open D.S., click on enter data icon, enter data in table, t in the x column and x in the y column. In the graph window, on the right side next to graph icon, pull down list and unclick connected lines. Click on the calculator in the graph window. Choose OK. On the calculator screen, under variables, choose model range, then enter data range, and enter equation, using * for multiplication and ^ for exponent.
This should produce a graph of your theoretical curve and also show your experimental data points.

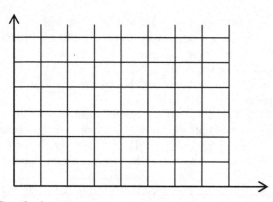

Conclusions
How well did your theoretical prediction agree with the observed motion? Discuss possible sources of error.

The tow rope was not exactly horizontal, as assumed in the theoretical analysis. Discuss the possible significance of this in your experiment. Use trig to determine the angle; don't guess. Discuss your results with the instructor.

Lab 9 Concept Quiz: Newton's Second Law; Rollerblade Platform
Suppose you are on a skateboard and initially you are at rest on a flat, level sidewalk. A rope is attached to your belt. Your friend who is standing nearby begins to pull on the rope. You are passive, just standing on your high quality, frictionless skateboard. The tension in the rope is 30 N and is maintained at this constant value.
1 What happens to you? A) nothing; B) you begin moving at a constant velocity; C) you begin moving towards your friend at a velocity that continually increases.
2 The tension in the rope means that: A) there is a 30 N pull on you, but no pull on your friend, and no net force on the rope; B) there is a 30 N pull on you, and also a 30 N pull on your friend, but no net force on the rope; C) there is a 30 N pull on you, a 30 N pull on your friend, and a 30 N net force on the rope; D) there is no pull on you or your friend, and no net force on the rope.
3 What happens to your friend? A) nothing; B) your friend begins moving at a constant velocity, the same velocity as you; C) your friend begins moving towards you at constant velocity; D) your friend begins to move in the same direction you are moving at a velocity that continually increases.
4 Is there is a friction force on your friend? A) yes; B) no.
   (Answer **5** only if you answered yes to **4**.)
5 Assume your friend has the same mass as you. The friction force on your friend is: A) 30 N, opposite the direction of her motion; B) 30 N, in the direction of motion; C) 60 N in the direction of motion; D) 60 N, opposite the direction of motion.

6 Draw complete free-body diagrams of: **a)** you; **b)** your friend; **c)** the rope; **d)** you, your friend, and the rope together.

a	b	c	d

Lab 10: Measurement and Uncertainty
Equipment: strips of paper with lines marked with times, small aluminum cylinder, electronic mass balance, digital calipers or micrometer

I Reaction Time
Measure your reaction time. A strip of paper is marked with lines and times in fractions of a second, from 0 to 0.20 s. While your partner holds the top of the paper, position your thumb and index finger on either side of the paper at the bottom at the zero mark. Your partner should release the paper without warning, and you should try to close your thumb and finger to catch it as quickly as possible. The farther the paper drops before you catch it, the greater your reaction time. Perform the experiment 10 times and record the time for each drop. Then switch roles and measure your partner's reaction time in the same way.

Name:

Time, s

Name:

Time, s

Suppose you are trying to decide who would make a better goalie on a soccer team. You want to choose someone who has the fastest reaction time. How would you decide? Compare and discuss your results. What would be a reasonable way to determine who has the quicker reaction time?

II Measurement and Uncertainty Theory

No measurement is ever exact. There is always some uncertainty in any measured value. For example, if you measure a sprinter's time for the 100 m dash, you may be able to measure the time to one tenth of a second or even to one thousandth of a second, depending on your measuring system. But even when your measurement is very precise, it is never exact. It is not possible to know that the time is *exactly* equal to some number, say *exactly* 10 s. Whenever we express the value of a measured quantity, it is important to give some idea of the uncertainty in that value. Sometimes it is enough to roughly indicate uncertainty through the number of decimal places we use to express the result. If we measure the sprinter's time to a tenth of a second, we might find $t = 10.2$ s, meaning that the time is within one-tenth of a second of 10.2 s. But if we measure the sprinter's time to the hundredth of a second, then we might measure it to be 10.24 s, meaning that our uncertainty is no more than one-hundredth of a second. We call the number of digits used the number of **significant figures**. A time of 10.2 s is measured to three significant figures, whereas a time of 10.24 s is measured to four significant figures. If we want to be more precise about just how uncertain we are, we can indicate our best estimate of the time, along with our uncertainty. For example, we might measure the sprinter's time to be 10.240 ± 0.004 s.

Errors in Measurement

Sometimes your measurements will be in error. The error could simply be a mistake, such as misreading the numbers on a scale. Or there could be something about the system or the measuring process that causes the error. Such errors are called **systematic errors**. For example, if you are using a spring scale to measure a force, your measurement might be consistently higher than the correct value because the scale has not been properly calibrated; the measuring instrument has caused the error. You can reduce the number of mistakes and systematic errors in your measurements by being careful, calibrating scales if feasible, repeating measurements, and by having your lab partner check your measurements.

Instrumental Uncertainty

Often the uncertainty in a measurement is determined by the smallest divisions on the scale of the measuring instrument. An appropriate choice of measuring instrument can reduce the uncertainty. For example, suppose that you are measuring the diameter of a ball bearing. If you use a ruler or meter stick, whose smallest divisions are millimeters, the best you can do is to measure the diameter to the nearest millimeter. However, with a micrometer you can measure the diameter to one-hundredth of a millimeter. Thus you might measure the ball's diameter with each of these instruments and obtain values of 3 mm and 3.18 mm, using the meter stick and micrometer respectively.

Random Errors and Average Values

Repeated measurement of a physical quantity often results in small fluctuations in the measured value. You should then use the average value of the individual measurements as your measured value. Doing so will tend to reduce the effect of small, random errors, which are present in individual measurements, but which tend to cancel out when averaged over a number of measurements. Even averaging a few such measurements will help to reduce error, and it will also allow you to estimate the uncertainty.

For example, suppose you fire balls from a spring launcher, and you want to know how far from the launcher a ball hits the ground. When you perform the experiment, you get somewhat different distances every time you fire the launcher. Suppose you make three measurements and get 80 cm, 81 cm, and 79 cm. The average of these three measurements is 80 cm, and all measurements are within 1 cm of 80 cm, so it is reasonable to express your result as:

$$x = 80 \text{ cm} \pm 1 \text{ cm}$$

Suppose you repeat the experiment a large number of times and find that almost all of the balls are within some distance Δx of an average value x_{av}. Then you could express your answer as

$$x = x_{av} \pm \Delta x$$

Percent Uncertainty

Percent uncertainty is usually a better measure of precision than absolute uncertainty. Suppose, for example, we measure the distance between two objects to be 15 cm, with an uncertainty of 0.6 cm. This is a 4% uncertainty: (0.6/15) × 100. But if we measure the distance between two other objects to be 2 cm with the same uncertainty of 0.6 cm, this is a 30% uncertainty (0.6/2) × 100, a much greater value. The 15 cm measurement is more precise in terms of percentage than the 2 cm measurement, though the absolute uncertainty of each measurement is the same.

Uncertainty in Computed Values

Often a measurement requires a calculation to obtain the final desired quantity. For example, if we want to measure the speed of a car, we may accomplish this by measuring the distance the car travels and the time it takes to cover that distance. Both measurements are uncertain, and so is the computed value of speed. The uncertainty in a computed value can be obtained by using **extreme values** of the measured quantities to compute the extreme values of the computed quantity. For example, if a car travels 100 m $\pm$ 1 m in a time interval of 10 s $\pm$ 0.5 s, its speed is between 9.4 m/s (99 m/10.5 s) and 10.6 m/s (101 m/9.5 s), and should be stated as 10.0 m/s $\pm$ 0.6 m/s.

Most of the computations in physics lab involve only multiplication, division, or raising to a power, but not addition, subtraction, or other operations. In such calculations, computation of uncertainty is simplified. When one of the measurements has a much higher percentage uncertainty than any of the others, the overall percentage uncertainty is the same as that of the highest uncertainty measurement. For example, in the previous example involving measuring the speed of a car, suppose that the distance is measured as 100 m $\pm$ 0.1 m (0.1% uncertainty) and the time is measured to be 10 s $\pm$ 0.5 s (5% uncertainty). The uncertainty in speed will be 5% (10 m/s ± 0.5 m/s), the same percent uncertainty as the time measurement, because that measurement's uncertainty is so much greater than the uncertainty in the distance measurement. This is called the **weakest link rule**.

If none of the percent uncertainties in measured values is much greater than the others, then you must either use the extreme value method, or if all of the % uncertainties are relatively small (say 5% or less), then the uncertainty of the computed value is approximately equal to the **sum** of the percentages of all the measured quantities that are multiplied or divided in the calculation. So, for the calculated speed of the car when distance is 100 m ± 1 m (1% uncertainty) and time is 10 s ± 0.5 s (5%), we simply add the two percentages: 5% + 1% = 6% to find the percent uncertainty in the speed: 10 m/s ± 0.6 m/s (6%).

Comparing Results of Measurement with Theory or with Other Experiments

Often we wish to measure a quantity and compare our measured value with a value predicted by theory. If the theoretical value is within the range of our measured values, then theory and experiment are consistent. If it is outside that range, then theory and experiment are not consistent.

Sometimes two different groups perform the same experiment, and we wish to determine if the results are in agreement. Agreement does not mean that the averages of the two sets of measurements are exactly equal, but only that there be an overlap between the two ranges of possible values found in the two experiments. For example, if one group measures x to be 34.7 cm ± 0.2 cm and a second group measures x to be 35.1 cm ± 0.5 cm, the two measurements are in agreement, because there is overlap in the two ranges of values (34.5 cm to 34.9 cm and 34.6 cm to 35.6 cm).

Summary

When you are making a measurement in lab:

a) Decide which factors affect your result most. Wherever possible, try to reduce the uncertainty of these factors; for example, try to reduce percent uncertainties by measuring longer distances or time intervals.

b) If one of the uncertainties (in %) is much larger than the others, you can ignore all the other uncertainties and use this value to express the percent uncertainty of the quantity you are calculating. Otherwise you must use one of the other methods described above.

c) Make a judgment about your results, taking into account the uncertainty.

III Measuring the Density of Aluminum

Density is defined as the mass per unit volume of a substance. Measure the density of aluminum, using the small cylindrical sample provided.

1 Use the digital balance to measure the sample's mass and its uncertainty.

Mass m = _____ g   Uncertainty in mass Δm = _____ g   % uncertainty _____

Labs and Lab Concept Quizzes

2 Use the digital calipers or micrometer to measure both the cylinder's height h and its uncertainty Δh and the cylinder's radius r and its uncertainty Δr.

Height $h =$ ____ cm $\Delta h =$ ____ cm % uncertainty = ____

Radius $r =$ ____ cm $\Delta r =$ ____ cm % uncertainty = ____

3 Compute the density.

4 Compute the uncertainty in your measurement of density. Begin by computing the % uncertainty. Show your work.

% uncertainty = ____   Uncertainty in density $\Delta \rho =$ ____

Your measured value of $\rho =$ ____ $\pm$ ____ g/cm^3
Density of pure aluminum = 2.70 g/cm^3. Density of aluminum alloys varies from 2.68 to 2.82 g/cm^3.

Lab 10 Concept Quiz: Measurement and Uncertainty

1 What is the percent uncertainty in the length measurement 2.35 ± 0.25 m?

2 Suppose that after a hike in the mountains, a friend asks how fast you walked. You recall that the trail was about 6 miles long and that it took between 2 and 3 hours. What is your average speed if you assume a time of 2.5 h?

3 Assume that the absolute uncertainty in your distance measurement is 0.1 mile. Estimate the relative (percentage) uncertainty in your distance measurement.

4 What is the absolute uncertainty in your time measurement?

5 What is its relative uncertainty?

6 Compare the relative uncertainties in time and distance. Which measurement is more accurate?

7 Can you use the weakest link rule?

Determine the relative and absolute uncertainties in the speed estimation.

Lab 11: Energy
Equipment: video camera, computer, LoggerPro software, meter stick, model roller coaster (strip of curved sheet metal mounted in plywood frame), Nakamura car (very low friction), two marbles of the same size, but very different mass, brass spring and support for vertical oscillation, masses and holder, loop-the-loop.

I Roller coasters and Energy
Determine whether mechanical energy is conserved for a car on a roller coaster.

1 $U_G \rightarrow K \rightarrow U_G$ Measure the height of the car's center of mass at the starting point where it is released from rest, and measure the height of the center of mass at the highest point on the other side when it instantaneously comes to rest. It is convenient to choose the height of the tabletop to be the zero of potential energy – to measure the height of the car above the tabletop. Use these measurements to determine the car's initial energy and final energy.

U_{Gi}	K_i	E_i	U_{Gf}	K_f	E_f	% difference, $E_f - E_i$

Conclusions

2 $U_G \rightarrow K$ Release the car from rest, with its center of mass 20 cm above the tabletop. Use the video camera to measure the speed of the car at the lowest point on the roller coaster. Measure the center of mass of the car above the tabletop at its lowest point. Here it is convenient to choose that lowest point as the zero of potential energy. Use these measurements to determine the car's initial energy (all potential) and final energy (all kinetic).

U_{Gi}	K_i	E_i	U_{Gf}	K_f	E_f	% difference, $E_f - E_i$

Conclusions

II Projectile Motion and Energy
1 Consider two balls of different mass, both given the same initial horizontal velocity off of a tabletop. Before performing the experiment, use your knowledge of projectile motion to predict which ball will travel the greater horizontal distance d from the edge of the table, or whether they will travel the same distance.
Prediction: Explain:

Experiment: Sweep the two balls together across the tabletop and off the edge, so as to give them the same horizontal initial velocity. Was your prediction verified?

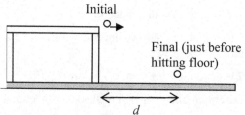

2 Use conservation of mechanical energy to relate the ball's final speed v_f just before hitting the floor to its initial speed v_i just after it leaves the tabletop and to the initial elevation y_i.
Does this relationship involve the mass of the ball?
Does this relationship tell you anything about the time to hit the floor or about the components of velocity?

III Mass Oscillating on a Spring

A mass hanging from a spring oscillates up and down. What are the three types of mechanical energy?
Describe how they vary throughout the motion.

1 Measure the force constant of a spring

Hang a 250 g mass from the brass spring with the mass at rest. Analyze the forces and determine the force constant of the spring.

2 Predict the lowest point

Suppose you start with the 250 g mass attached to the spring, with the spring initially unstretched and the mass at rest, and then release the mass. Use energy analysis to predict the lowest point reached by the mass when the spring is stretched the most.
Draw initial and final sketches of the system.

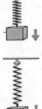

What assumption do you make in your analysis?

For simplicity it's a good idea to define the zero of both kinds of potential energy to be at the high point. What then is the value of the total mechanical energy at that point?

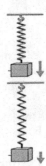

Predicted distance that mass drops from high point to low point:

Once you have made your prediction, perform the experiment. Rather than trying to locate the center of mass of the system, which can be tricky, it's OK to just measure a fixed point. The top of the mass works well. Measured distance that mass drops from high point to low point:

IV Circular motion & energy

If a ball starts high enough on the loop-the-loop, it will travel all the way around the vertical circle. We can use conservation of mechanical energy to determine how high above the top of the circle the ball must start. Draw a free body diagram of the ball at the top of the circle. At that point it is moving. If it weren't it would simply fall straight down.

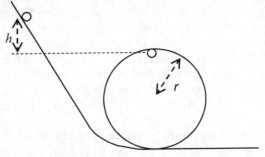

Analyze your FBD using Newton's second law to find an expression for the minimum speed the ball must have at the top of the circle. Then use conservation of mechanical energy to show that the ball's minimum height h above the top of the circle equals ½ r, where r is the radius of the circle. (This analysis ignores the rotational kinetic energy of the ball. If we take it into account the result changes to 0.7 r.) Test your prediction.

V Energy Skate Park Simulation

1 Open the PhET simulation Energy Skate Park. Open the bar graph to see the values of kinetic and potential energy change as the skater moves. Set speed to slow initially so you can easily see what is happening. Do kinetic and potential energies change?   Does the skater's total mechanical energy change? Adjust the zero of potential energy to see how this changes things.

2 Turn on friction. The initial default setting is for no friction. How does this change the motion? The total energy in the bar graph now includes thermal energy. Is mechanical energy still conserved?   What happens to the total mechanical energy if you leave it running long enough?

3 Add track to the run to create your own design, with more interesting motion. Sketch the run you create.

Lab 11 Concept Quiz: Energy

1 Energy is: A) always conserved; B) sometimes conserved; C) never conserved.
2 Mechanical energy is: A) always conserved; B) sometimes conserved; C) never conserved.
For questions **3 – 5**, consider the loop-the-loop demonstration, in which we release a ball from a point on a track well above the top of a vertical circular section of track.
3 To find the speed of a ball at the top of the loop-the-loop, can one use conservation of mechanical energy? A) yes, if friction is negligible; B) yes, whether or not friction is negligible; C) no.
4 A) You can't use conservation of mechanical energy to find the speed of the ball at the top of the loop-the-loop; B) You can use conservation of mechanical energy to find the speed, and to do so you must know the mass of the ball; C) You can use conservation of mechanical energy to find the speed, and to do so you need not know the mass of the ball.
5 A) You can't use conservation of mechanical energy to find the speed of the ball at the top of the loop-the-loop; B) you can use conservation of mechanical energy to find the speed, and to do so you must know the radius of the loop; C) you can use conservation of mechanical energy to find the speed, and to do so you must know only the height of the starting point above the top of the loop; D) you can use conservation of mechanical energy to find the speed, and to do so you must know only the height of the starting point above the top of the loop and the radius of the

loop; E) you can use conservation of mechanical energy to find the speed, and to do so you must know the height of the starting point above the top of the loop, the radius of the loop, and the mass of the ball.

For questions 6 – 8, consider a projectile with known Cartesian coordinates throughout its trajectory and known velocity at a single point on the trajectory and known time at which it is at that point. We assume air resistance is negligible.

6 Can we use conservation of mechanical energy to find the velocity of the projectile at any point? A) yes, if we know the mass of the projectile; B) yes, whether or not we know the mass of the projectile; C) no.

7 Can we use conservation of mechanical energy to find the speed of the projectile at any point? A) yes, if we know the mass of the projectile; B) yes, whether or not we know the mass of the projectile; C) no.

8 Can we use conservation of mechanical energy to find the time at which the projectile is at any point? A) yes, if we know the mass of the projectile; B) yes, whether or not we know the mass of the projectile; C) no.

Lab 12: Static Equilibrium

Equipment: plastic meter stick, 1 cm thick plywood block, 2 universal Pasco table clamps, assorted cylindrical brass masses, 2 small table stands, 2 small pulleys and 2 small 90° clamps to attach them to stands, string, scissors, tape, two small mass hangers and masses, small protractor, 2 playing cards, board to be used as an inclined plane.

I Balancing a Meter Stick

Turn the 1 cm plywood block on edge and secure at the edge of the table with table clamps on either side. The block will serve as a support and potential pivot point for the meter stick. Balance the meter stick on the plywood edge. It should balance when the 50 cm mark is directly over the center of the plywood edge.

1 Place two 100 g masses on top of the meter stick on opposite sides of the midpoint in such a way that the meter stick is balanced. Record their distances from the midpoint: $r_1 =$ _____ $r_2 =$ _____ Draw a complete FBD of the meter stick. Compute and compare positive and negative torques with respect to an axis through the center of the meter stick. To simplify calculation, use $g = 10$ m/s^2.

FBD:

Positive torque _____   Negative torque _____

2 Place a 100 g mass and a 200 g mass on top of the meter stick on opposite sides of the midpoint in such a way that the meter stick is balanced. Record their distances from the midpoint: $r_1 = $ _____ $r_2 = $ _____ Draw a complete FBD of the meter stick. Compute and compare positive and negative torques.

FBD:

Positive torque _____ Negative torque _____

3 Place a 50 g mass and a 100 g mass on top of the meter stick on one side of the midpoint and a 100 g mass and a 200 g mass on the other side, in such a way that the meter stick is balanced. Record their distances from the midpoint.
Left side: $m = 50$ g, $r = $ _____ $m = 100$ g, $r = $ _____ ; Right side: $m = 100$ g, $r = $ _____ $m = 200$ g, $r = $ _____. Draw a complete FBD of the meter stick. Compute and compare the sum of the positive torques and the sum of the negative torques.

FBD:

Sum of positive torques _____ Sum of negative torques _____

4 Place the 30 cm mark on the meter stick directly over the plywood edge so that the weight of the meter stick, acting at its center of gravity, provides a torque with respect to an axis through the 30 cm pivot point. Place on top of the meter stick a mass, positioned so that the meter stick balances. Draw a complete FBD of the meter stick. Make measurements and use them to determine the mass of the meter stick.

FBD:

Mass of meter stick _____

5 Tape a string to the left side of the meter stick at the 30 cm mark. Pass the string over a pulley and attach it to a 55 g mass, so that the mass is supported by the string. Adjust the position of the pulley so that the string makes an angle of 45° with the horizontal. You will balance the torque on the meter stick produced by this string with a torque produced by a string on the opposite side, attached at the 90 cm mark and supporting an equal mass, but before doing so, draw a FBD, and use the condition for rotational equilibrium to predict the angle θ_2.

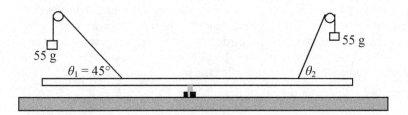

FBD:

Predicted value of θ_2 _____   Measured value of θ_2 _____

II Balancing Playing Cards

1 Two playing cards leaning against each other can be balanced to form a tent, as shown in the figure, so long as the angle ϕ is not too small. The value of the minimum angle depends on the coefficient of static friction. Find an expression for the minimum angle as a function of the coefficient μ_s. The result is independent of both the length and weight of a playing card. Hint: You will need two FBDs, one to relate the normal force to the weight of a card and the other, with a wise choice of axis, to relate the coefficient to the angle. Check your result with the instructor before proceeding.

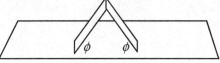

FBD 1:   FBD 2:

Equation for minimum angle: tan ϕ = _____

2 Determine the coefficient of static friction between a playing card and a paper surface by laying the playing card on a sheet of paper taped to a board. Gradually raise one end of the board until the playing card begins to slide along the paper. Measure the tangent of the angle θ the incline makes with the horizontal as the sliding begins. Show that the value of that tangent equals the coefficient of static friction between paper and playing card. Check your result with the instructor before proceeding. Use your measurement of tan θ as the value of μ_s.

FBD: tan θ = μ_s = _____

3 Use the results of **1** and **2** to predict the minimum angle ϕ of the playing cards to balance on a paper surface. Then perform the experiment and measure that angle. To simplify the experiment, tape the top edges of the two playing cards together.

Predicted minimum angle _____   Measured minimum angle _____

13

Quizzes and Tests

"I'm not telling you it is going to be easy; I'm telling you it's going to be worth it."
Art Williams

FIRST SEMESTER ONLY (2nd semester tests available on TIP website)
Quiz 1
Question (4 points)
The dots in this figure show the positions of two objects, A & B, at equal time intervals. Sketch on a single graph the position x vs. t for each of the objects.

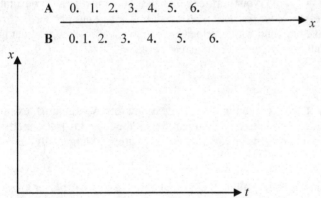

Are the positions of the two objects on the x axis ever equal? If so, when?

Problem (6 points)

Find the resultant force for the forces shown in the figure.

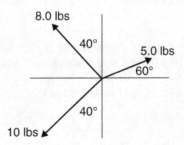

Quiz 2
Question (4 points)
The graph below shows the position of an object moving along the x axis as a function of time t. Sketch a graph of the object's velocity v_x vs. t.

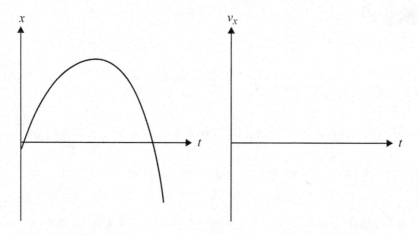

Problem (6 points)
As you are sitting in this room, your partner walks past you, moving at a constant velocity of 3.0 m/s, east. Your partner tosses a ball upward. Your partner observes the ball leave her hand with a velocity of 6.0 m/s, vertically up. What is the velocity of the ball relative to you as it leaves her hand?

Quiz 3
Questions (4 points)
1 Your small, compact car is parked in the LMU parking lot. A large truck exiting a parking space collides with your car. Compared with the force the truck exerts on your car, the force your car exerts on the truck is: **i)** smaller; **ii)** larger; **iii)** the same.

2 A soccer ball is initially rolling along the ground, moving East. A player kicks the ball, briefly exerting a force on it, directed North. After the kick, the ball's component of velocity toward the East is: **i)** less than before the kick; **ii)** greater than before the kick; **iii)** the same as before the kick.

Problem (6 points)
A skater of mass 80 kg skates on frictionless ice toward you, moving at a constant velocity of 5.0 m/s. You make contact with the skater for 2.0 s, exerting on the skater a force of 160 N, directed opposite to the skater's velocity. What is the velocity of the skater at the end of the 2.0 s during which you apply the force?

Quiz 4
Questions (4 points)
1 A crate is on a flatbed truck that is traveling along a flat, level road. Nothing touches the crate except the horizontal surface of the truck bed. As the driver approaches a stoplight, she applies the brakes. As the brakes are applied: **i)** there is a friction force exerted on the crate by the truck in the direction of the truck's motion; **ii)** there is a friction force exerted on the crate by the truck in the direction opposite the truck's motion; **iii)** there is no friction exerted on the crate by the truck.

2 In the demonstration in class, a sheet of paper was jerked out from beneath a book, the book remaining at rest. The paper moved to the right. The book exerted on the paper: **i)** a friction force to the right; **ii)** a friction force to the left; **iii)** no friction force.

Problem (6 points)
An iPhone of mass 150 g is pushed against a wall by a force **P** of magnitude 3.0 N as shown. If the coefficient of kinetic friction between phone and wall is 0.30, what is the phone's acceleration?

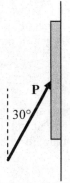

Quiz 5
Question (4 points)
You are on an amusement park ride at the top of a loop-the-loop, a vertical circle, riding upside down. At that instant indicate which of the following directions A, B, C, D, or E, shown in the figure, is the direction of the following vectors.
Your velocity ____   Your acceleration ____
The resultant force on you ____

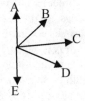

Would any of these quantities be in the same direction at an instant when you were on the side of the loop, where the track is vertical?

Problem (6 points)
In the situation described in the question, at the top of the loop, suppose your mass is 50 kg, the radius of the loop is 20 m, and you are moving at a speed of 20 m/s. Identify all of the forces acting on your body and the source of these forces, and find the magnitude and direction of the forces.

Quiz 6
Questions (4 points)
A car has an initial kinetic energy of 100,000 J, and then it stops as the brakes are applied.
1 What is the net work done on the car as it is stopping?
i) 100,000 J; **ii)** –100,000 J; **iii)** 0; **iv)** 200,000 J; **v)** – 200,00 J.

2 You drive at a constant speed of 70 mph downhill on a freeway. Is your car's mechanical energy conserved? **i)** yes; **ii)** no.

3 Considering all types of energy, mechanical or otherwise, is total energy conserved? **i)** yes; **ii)** no.

Problem (6 points)
Suppose you are designing a new slide for a playground. To offer more excitement, you want to increase the maximum speed of a child at the bottom. How high should the new slide be, in order to double the speed at the bottom, if the vertical height of the old slide is 1.5 m? Assume friction to be negligible.

Quiz 7
Questions (4 points)

1 The body shown in this figure is subject only to the two forces shown. Is the body in static equilibrium?
i) yes; **ii)** no; **iii)** depends on the axis.

2 A painter stands on a board that does not move. The sum of the torques about an axis through A is zero. The sum of the torques about an axis through B is: **i)** positive; **ii)** negative; **iii)** zero.

Problem (6 points)
You lean a 2 m long uniform board against a vertical wall in a room with smooth, frictionless walls. What is the maximum angle θ between the wall and the board at which the board will not slide downward if the coefficient of static friction between the board and the floor is 0.5? To solve:
a) Choose a free body and draw an appropriate FBD that will allow you to solve this problem. What are the general equations you will apply to the free body?
b) The board weighs 10 N. What is the value of the other forces on the board?
c) Choose an axis of rotation and find the torque for each force. Each nonzero torque will be expressed as a function of θ.
d) What is the value of θ at which the board begins to slide?

PHYS 253 Test 1 (chapters 1 – 4)
Questions (4 points each)

1 You are walking across campus at a constant speed of 3 m/s and a friend, who is initially 40 m in front of you, sees you and runs toward you at a constant speed of 5 m/s. Five seconds later you and your friend meet. Draw a sequence of dots to represent your position at 1 s intervals from 0 to 5 s and a sequence of X's to represent your friend's position for these times. Let the positive x axis be in the direction of your motion, with you starting at the origin at $t = 0$.

2 For the motion described in **1**, graph your position x vs. time t as a solid line below, and draw on the same graph a sketch of your friend's position vs. time as a dashed line.

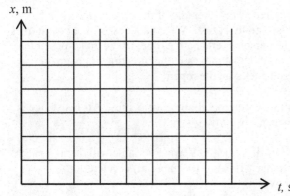

3 For the motion described in 1, graph below your velocity v_x vs. t as a solid line, and draw on the same graph a sketch of your friend's velocity v_x vs. t as a dashed line.

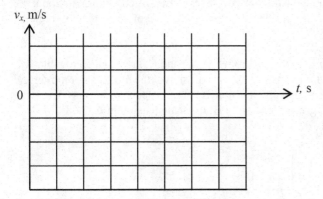

4 For the motion described in **1**, what was the velocity of your friend relative to you?

5 A 30 N force acts on an object and a 40 N force, directed perpendicular to the 30 N force also acts on the object. The combination of the two forces is equivalent to a single force of magnitude: **i)** 10 N; **ii)** 20 N; **iii)** 50 N; **iv)** 70 N; **v)** can't say without more information.

6 You are walking down the hall toward this classroom in the positive y direction, and then you reverse your motion and begin walking in the opposite direction. What is the direction of your acceleration? **i)** there is no acceleration; **ii)** positive y direction; **iii)** negative y direction; **iv)** positive x direction; **v)** negative x direction.

7 You are in a boat moving east at 20 m/s when you see another boat, which, relative to you, is moving with a velocity of 20 m/s west. What is the velocity of the other boat relative to the Earth? **i)** 20 m/s, W; **ii)** 20 m/s, N; **iii)** 20 m/s, S; **iv)** 20 m/s, E; **v)** it's not moving.

8 One end of a rope is tied to a tree. The other end of the rope is pulled on by someone in a small boat that is initially at rest. What happens? **i)** the boat moves in the direction of the force the person exerts on the rope; **ii)** the boat does not move because forces cancel; **iii)** the boat moves in the direction opposite the direction of the force the person exerts on the rope.

9 Several darts are thrown, all with horizontal initial velocities. Air resistance is negligible. Draw a FBD of one of the darts.

10 During the flights of the darts described in **9**, what variable will distinguish the trajectory of one dart from another? **i)** mass; **ii)** shape; **iii)** acceleration; **iv)** velocity; **v)** none of the above.

11 You are in your Ferrari, stopped at a light. You decide to see how fast you can accelerate once the light turns green. As you do, you feel yourself pushed back in your seat. What force is responsible for your backward motion? **i)** gravitational force; **ii)** force of the road; **iii)** acceleration force; **iv)** force of the engine; **v)** none of the above.

12 In a special three team tug of war, 3 ropes, A, B, and C, are attached to a block and pulled in 3 different directions. At the instant shown the block is not moving, but all ropes are under tension. Is the tension greater in rope A or in rope B? **i)** greater tension in A; **ii)** greater tension in B; **iii)** same tension in A and B.

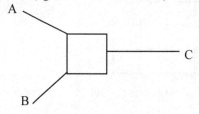

Problems (13 points each)   **Show all your work.**
1 A basketball player is positioned 13 m away from a point directly under the basket when he makes a shot, imparting to the basketball an initial velocity with a horizontal component of 10 m/s and a vertical component of 10 m/s. Find the ball's components of velocity when it goes through the basket.

2 A car of mass 1,000 kg is initially traveling east on a level road at a speed of 25 m/s, and then, during a 6.0 s time interval, turns left and begins moving north at 35 m/s. What is the average horizontal force acting on the car during the turn?

3 A goalie kicks a soccer ball almost the entire length of a soccer field. Make reasonable estimates about the ball, the field, and the distance the goalie's foot moves while it is in contact with the ball, and use these estimates to estimate:
    a) the ball's initial velocity as it leaves the foot;
    b) the acceleration of the ball as it is being kicked;
    c) the force exerted by the goalie's foot on the ball.

4 A box of mass 5.0 kg is being pushed up a frictionless 30° incline by a horizontal 40 N force. How long will it take to move the box 1.0 m, starting from rest?

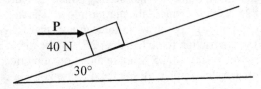

PHYS 253 Test 2 (chapters 4, 5, 6, and 8)
Questions (4 points each)

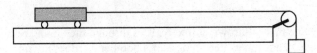

1 A 1 kg model car is moving to the right on a horizontal track at constant velocity. Attached to the car is a horizontal string that passes over a pulley. The other end of the string is attached to a hanging mass that weighs 3.0 N. The friction force on the car is: **i)** 0 N; **ii)** < 3.0 N, but not 0; **iii)** 3.0 N; **iv)** > 3.0 N.

2 A force in the positive y direction acts on an object that is initially moving in the positive x direction. What happens? **i)** object immediately begins moving in the positive y direction; **ii)** object moves at a constant velocity with components in both the x and y directions; **iii)** object continues to move in the x direction;
iv) object has velocity components in both the x and y directions & both continually increase; **v)** object begins to have a component of velocity in the y direction that continually increases, while having a constant velocity component in the x direction.

3 A car is parked on a hill, heading uphill. **i)** there is no friction force on the car, but there is a normal force equal to the weight of the car; **ii)** there is a friction force directed down the hill and a normal force equal to the weight; **iii)** there is a friction force directed up the hill and a normal force equal to the weight; **iv)** there is a friction force directed up the hill and a normal force smaller than the car's weight.

4 You are standing on a merry-go-round not touching anything but the horizontal floor of the ride with your feet. Is there a friction force on your feet? **i)** no; **ii)** yes, directed outward away from the center of the ride; **iii)** yes, directed inward toward the center of the ride.

5 In the Rollerblade platform experiment, when Jeffrey pulled Chandler, pulling on the rope with a force of 25 N, what was the resultant force on the rope? **i)** 25 N; **ii)** a little less than 25 N; **iii)** 0.

6 In the Rollerblade platform experiment, when Jeffrey pulled Chandler, pulling on the rope with a force of 25 N, what force, if any, was exerted by the rope on Jeffrey? **i)** 25 N; **ii)** a little less than 25 N; **iii)** 0.

7 In the Rollerblade platform experiment, when Jeffrey pulled Chandler, pulling on the rope with a force of 25 N, which caused the platform to accelerate, was there a friction force on Jeffrey? **i)** no; **ii)** yes, a friction force greater than 25 N in the direction of motion; **iii)** yes, a friction force less than 25 N in the direction of motion; **iv)** yes, a friction force equal to 25 N in the direction of motion; **v)** yes, a friction force greater than 25 N opposite the direction of motion.

8 Planet X has half the mass of Earth and half the radius of Earth. If you weigh 600 N on Earth, how much would you weigh on this planet? **i)** 300 N; **ii)** 600 N; **iii)** 900 N; **iv)** 1200 N.

9 A boomerang is a device that, if thrown correctly, flies through the air and returns to the person throwing it. Thus the center of mass of a boomerang does not obey simple projectile motion. Suppose a boomerang were thrown on the moon, where there is no atmosphere. Would it return or would it follow simple projectile motion? **i)** it would return; **ii)** it would follow projectile motion. **Explain**.

10 A crate is on the flat bed of a truck and does not slide along the surface as the truck makes a right turn at constant speed. There is no force on the box, other than perhaps friction, that prevents the box from sliding along the truck bed. The friction force of the surface on the crate is: **i)** zero; **ii)** nonzero, directed to the left; **iii)** nonzero, directed toward the front of the truck; **iv)** nonzero, directed toward the back of the truck; **v)** nonzero, directed to the right; **vi)** nonzero, directed to the left.

11 A skier on a ski jump accelerates down a ramp that curves upward at the end so that the sker is launched through the air. At which point, if any, is the normal force exerted by the ramp on the skier equal to the skier's weight? **i)** near the top of the ramp; **ii)** at the lowest point on the ramp where the shape of the ramp is a circular arc; **iii)** at the right end of the ramp; **iv)** after leaving the ramp; **v)** nowhere.

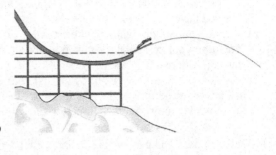

12 You whirl a ball attached to a string in a horizontal plane overhead. The string breaks at the instant the ball's velocity points in the direction of a telephone pole. Will the ball hit the pole? **i)** yes; **ii)** no. Sketch the path of the ball as seen from above. Include the pole in your sketch.

Extra Credit
13 You find yourself in the front of a canoe without a paddle, at rest on a still lake, just a few meters from shore, just out of reach of your friend who is on shore in front of the canoe and wants to pull the canoe to shore. What can you do to enable your friend to reach the canoe? **i)** move to the back of the canoe; **ii)** jump up and down; **iii)** there is nothing you can do, other than get out of the canoe and swim. Explain.

Problems (13 points each)   **Show all your work.**

1 Two of the Rollerblade platforms used in the last lab are positioned one in front of the other. The carts each have a mass of 5 kg. The front cart carries a rider of mass 55 kg, and the rear cart carries a rider of mass 75 kg. The carts initially move forward together at a speed of 3.0 m/s. They move together initially because in the rear cart the rider's feet are holding onto the front cart. Then the rider in the back cart pushes the front cart away, causing it to move forward with a velocity of 7.0 m/s in the forward direction. What is the velocity of the back cart after the shove?

2 You are on a small asteroid of mass 1.0×10^{17} kg and radius 2.0×10^4 m. Suppose you have a mass of 100 kg.
  a) What is your weight while on the asteroid?
  b) You drop your cell phone. How long does it take for the phone to fall 1 m, starting from rest?
  c) You decide to amuse yourself by putting your cell phone into a circular orbit just above the surface of the asteroid. What is the magnitude and direction of the initial velocity you must impart to the cell phone to accomplish this?

3 A child of mass 20 kg swings back and forth on a playground swing. At some instant, the swing's chains make an angle of 30° with the vertical and the child is at the highest point in her swing, with an instantaneous velocity of zero. At that instant, find: **a)** the instantaneous acceleration of the child; **b)** the tension in each of the swing's two chains.

4 You exert a push **P**, directed at an angle of 40° with the horizontal, on your cell phone of mass 150 g, causing it to move at constant velocity along a horizontal surface, with a coefficient of kinetic friction between phone and surface of 0.30. Find the magnitude of the force **P**.

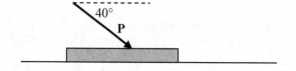

Appendix 1: The Lawson Classroom Test of Scientific Reasoning Ability

1. Suppose you are given two clay balls of equal size and shape. The two clay balls also weigh the same. One ball is flattened into a pancake-shaped piece. *Which of these statements is correct?*
    a. The pancake-shaped piece weighs more than the ball
    b. The two pieces still weigh the same
    c. The ball weighs more than the pancake-shaped piece

2. *because*
    a. the flattened piece covers a larger area.
    b. the ball pushes down more on one spot.
    c. when something is flattened it loses weight.
    d. clay has not been added or taken away.
    e. when something is flattened it gains weight.

3. To the right are drawings of two cylinders filled to the same level with water. The cylinders are identical in size and shape.

    Also shown at the right are two marbles, one glass and one steel. The marbles are the same size but the steel one is much heavier than the glass one.

    When the glass marble is put into Cylinder 1 it sinks to the bottom and the water level rises to the 6th mark. *If we put the steel marble into Cylinder 2, the water will rise*
    a. to the same level as it did in Cylinder 1
    b. to a higher level than it did in Cylinder 1
    c. to a lower level than it did in Cylinder 1

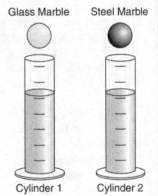

4. *because*
    a. the steel marble will sink faster.
    b. the marbles are made of different materials.
    c. the steel marble is heavier than the glass marble.
    d. the glass marble creates less pressure.
    e. the marbles are the same size.

5. To the right are drawings of a wide
and a narrow cylinder. The cylinders
have equally spaced marks on them.
Water is poured into the wide cylinder
up to the 4th mark (see A). This water
rises to the 6th mark when poured into
the narrow cylinder (see B).

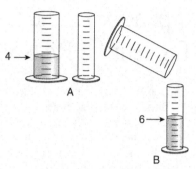

Both cylinders are emptied (not
shown) and water is poured into the
wide cylinder up to the 6th mark.
*How high would this water rise if it were
poured into the empty narrow cylinder?*
    a. to about 8
    b. to about 9
    c. to about 10
    d. to about 12
    e. none of these answers is correct

6. *because*
    a. the answer cannot be determined with the information given.
    b. it went up 2 more before, so it will go up 2 more again.
    c. it goes up 3 in the narrow for every 2 in the wide.
    d. the second cylinder is narrower.
    e. one must actually pour the water and observe to find out.

7. Water is now poured into the narrow cylinder (described in Item 5 above) up to
the 11th mark. *How high would this water rise if it were poured into the empty
wide cylinder?*
    a. to about 7 1/2
    b. to about 9
    c. to about 8
    d. to about 7 1/3
    e. none of these answers is correct

8. *because*
    a. the ratios must stay the same.
    b. one must actually pour the water and observe to find out.
    c. the answer cannot be determined with the information given.
    d. it was 2 less before so it will be 2 less again.
    e. you subtract 2 from the wide for every 3 from the narrow.

9. At the right are drawings of three strings hanging from a bar. The three strings have metal weights attached to their ends. String 1 and String 3 are the same length. String 2 is shorter. A 10 unit weight is attached to the end of String 1. A 10 unit weight is also attached to the end of String 2. A 5 unit weight is attached to the end of String 3. The strings (and attached weights) can be swung back and forth and the time it takes to make a swing can be timed. Suppose you want to find out whether the length of the string has an effect on the time it takes to swing back and forth. *Which strings would you use to find out?*

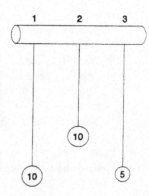

    a. only one string
    b. all three strings
    c. 2 and 3
    d. 1 and 3
    e. 1 and 2

10. *because*
    a. you must use the longest strings.
    b. you must compare strings with both light and heavy weights.
    c. only the lengths differ.
    d. to make all possible comparisons.
    e. the weights differ.

11. Twenty fruit flies are placed in each of four glass tubes. The tubes are sealed. Tubes I and II are partially covered with black paper; Tubes III and IV are not covered. The tubes are placed as shown. Then they are exposed to red light for five minutes. The number of flies in the uncovered part of each tube is shown in the drawing.

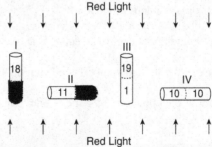

This experiment shows that flies respond to (respond means move to or away from):
    a. red light but not gravity
    b. gravity but not red light
    c. both red light and gravity
    d. neither red light nor gravity

12. *because*
    a. most flies are in the upper end of Tube III but spread about evenly in Tube II.
    b. most flies did not go to the bottom of Tubes I and III.
    c. the flies need light to see and must fly against gravity.
    d. the majority of flies are in the upper ends and in the lighted ends of the tubes.
    e. some flies are in both ends of each tube.

13. In a second experiment, a different kind of fly and blue light were used. The results are shown in the drawing.

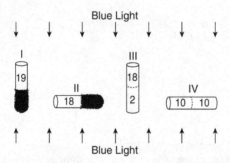

    These data show that these flies respond to (respond means move to or away from):
    a. blue light but not gravity
    b. gravity but not blue light
    c. both blue light and gravity
    d. neither blue light nor gravity

14. *because*
    a. some flies are in both ends of each tube.
    b. the flies need light to see and must fly against gravity.
    c. the flies are spread about evenly in Tube IV and in the upper end of Tube III.
    d. most flies are in the lighted end of Tube II but do not go down in Tubes I and III.
    e. most flies are in the upper end of Tube I and the lighted end of Tube II.

The Lawson Classroom Test of Scientific Reasoning Ability

15. Six square pieces of wood are put into a cloth bag and mixed about. The six pieces are identical in size and shape, however, three pieces are red and three are yellow. Suppose someone reaches into the bag (without looking) and pulls out one piece. *What are the chances that the piece is red?*
    a. 1 chance out of 6
    b. 1 chance out of 3
    c. 1 chance out of 2
    d. 1 chance out of 1
    e. cannot be determined

16. *because*
    a. 3 out of 6 pieces are red.
    b. there is no way to tell which piece will be picked.
    c. only 1 piece of the 6 in the bag is picked.
    d. all 6 pieces are identical in size and shape.
    e. only 1 red piece can be picked out of the 3 red pieces.

17. Three red square pieces of wood, four yellow square pieces, and five blue square pieces are put into a cloth bag. Four red round pieces, two yellow round pieces, and three blue round pieces are also put into the bag. All the pieces are then mixed about. Suppose someone reaches into the bag (without looking and without feeling for a particular shape piece) and pulls out one piece.

    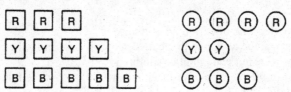

    What are the chances that the piece is a red round or blue round piece?
    a. cannot be determined
    b. 1 chance out of 3
    c. 1 chance out of 21
    d. 15 chances out of 21
    e. 1 chance out of 2

18. *because*
    a. 1 of the 2 shapes is round.
    b. 15 of the 21 pieces are red or blue.
    c. there is no way to tell which piece will be picked.
    d. only 1 of the 21 pieces is picked out of the bag.
    e. 1 of every 3 pieces is a red or blue round piece.

19. Farmer Brown was observing the mice that live in his field. He discovered that all of them were either fat or thin. Also, all of them had either black tails or white tails. This made him wonder if there might be a link between the size of the mice and the color of their tails. So he captured all of the mice in one part of his field and observed them. Below are the mice that he captured.

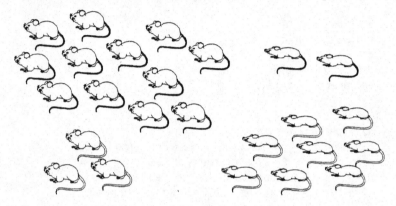

Do you think there is a link between the size of the mice and the color of their tails?

    a. appears to be a link
    b. appears not to be a link
    c. cannot make a reasonable guess

20. *because*
    a. there are some of each kind of mouse.
    b. there may be a genetic link between mouse size and tail color.
    c. there were not enough mice captured.
    d. most of the fat mice have black tails while most of the thin mice have white tails.
    e. as the mice grew fatter, their tails became darker.

21. The figure below at the left shows a drinking glass and a burning birthday candle stuck in a small piece of clay standing in a pan of water. When the glass is turned upside down, put over the candle, and placed in the water, the candle quickly goes out and water rushes up into the glass (as shown at the right).

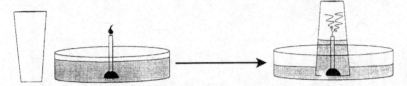

This observation raises an interesting question: Why does the water rush up into the glass?

Here is a possible explanation. The flame converts oxygen into carbon dioxide. Because oxygen does not dissolve rapidly into water but carbon dioxide does, the newly-formed carbon dioxide dissolves rapidly into the water, lowering the air pressure inside the glass.

Suppose you have the materials mentioned above plus some matches and some dry ice (dry ice is frozen carbon dioxide). *Using some or all of the materials, how could you test this possible explanation?*

    a. Saturate the water with carbon dioxide and redo the experiment noting the amount of water rise.
    b. The water rises because oxygen is consumed, so redo the experiment in exactly the same way to show water rise due to oxygen loss.
    c. Conduct a controlled experiment varying only the number of candles to see if that makes a difference.
    d. Suction is responsible for the water rise, so put a balloon over the top of an open-ended cylinder and place the cylinder over the burning candle.
    e. Redo the experiment, but make sure it is controlled by holding all independent variables constant; then measure the amount of water rise.

22. What result of your test (mentioned in #21 above) would show that your explanation is probably wrong?

    a. The water rises the same as it did before.
    b. The water rises less than it did before.
    c. The balloon expands out.
    d. The balloon is sucked in.

23. A student put a drop of blood on a microscope slide and then looked at the blood under a microscope. As you can see in the diagram below, the magnified red blood cells look like little round balls. After adding a few drops of salt water to the drop of blood, the student noticed that the cells appeared to become smaller.

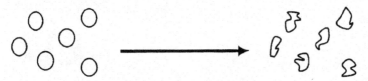

Magnified Red Blood Cells   After Adding Salt Water

This observation raises an interesting question: Why do the red blood cells appear smaller?

Here are two possible explanations: I. Salt ions (Na+ and Cl-) push on the cell membranes and make the cells appear smaller. II. Water molecules are attracted to the salt ions so the water molecules move out of the cells and leave the cells smaller.

To test these explanations, the student used some salt water, a very accurate weighing device, and some water-filled plastic bags, and assumed the plastic behaves just like red-blood-cell membranes. The experiment involved carefully weighing a water-filled bag in a salt solution for ten minutes and then reweighing the bag.

What result of the experiment would best show that explanation I is probably wrong?
  a. the bag loses weight
  b. the bag weighs the same
  c. the bag appears smaller

24. *What result of the experiment would best show that explanation II is probably wrong?*
  a. the bag loses weight
  b. the bag weighs the same
  c. the bag appears smaller

Appendix 2: A Guide to Learning Physics

The First Week – Getting to Know You

During the first week of Thinking in Physics, I will explain the nature of this course, its emphasis on both conceptual understanding and effective problem solving, and some of the reasons for the methods I use. I will also use this time to get to know you through various assessments, surveys, thinking journals, and personal interviews. My hope is that you are ready to commit to the work required to achieve the deep level of understanding that is my goal for you in this course. If so, I will do everything that I can to help you.

Neuroscientists and cognitive psychologists are discovering that the brain is malleable, and this is borne out in the experience of Thinking in Physics students. An essential part of Thinking in Physics (TIP) is recognizing and addressing students' scientific reasoning skills because these skills are very important for learning physics. TIP develops reasoning skills. Fig. 1 shows the average scientific reasoning skills of recent TIP students at the beginning and at the end of a recent TIP course. Fig. 2 shows the average of students' understanding of physics concepts at the beginning and end of the course. The final class average was 70% understanding of physics concepts. While this might seem low, it turns out that for a starting average below 30%, this final average is among the highest achieved in any college physics class in the country.

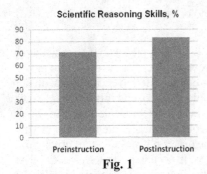
Fig. 1

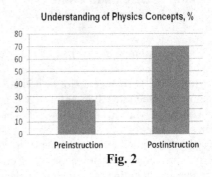
Fig. 2

Fig. 3 shows individual postinstruction results. Over half of all TIP students ended the course with at least a good understanding of physics concepts (meaning they understood at least 70% of the concepts), and one out of three students ended with an excellent understanding of physics concepts, (80% or more of concepts). On the other hand, one out of seven students ended the course understanding fewer than 50% of the physics concepts. You can certainly do very well in the Thinking in Physics course, but you will have to make a strong effort to be successful in learning physics.

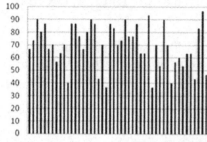

Fig. 3 At the end of a recent TIP course, the percentage of physics concepts understood by individual students ranged from 37% to 97%.

Fig. 3

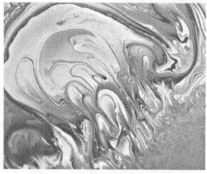

Seeing in Black and White
(black and white photo of oil film)

Seeing in Color
(color version of the same photo)

You may come to this course with a strong interest in physics. If so, you are fortunate because this will make the hard work enjoyable. Sadly, this is not the case for most life science majors who take this course. However, if you are open to developing some interest in physics, it is possible to do so, and this will likely make the work somewhat easier for you. One of the best ways to develop such an interest is to begin to notice how physics concepts and laws relate to things in your everyday life. For example, concepts in first semester physics relate to driving a car, playing sports, or just walking, and concepts in second semester physics relate to dangers of lightning and household electricity, how glasses or contacts help you see, and rainbows or other rarer optical phenomena. The physics perspective is a new way for you to view the world. Suppose you were completely color blind – not the common red-green kind of color blind, but rather the very rare condition in which you see the world only in black and white and grays. Then imagine that you could suddenly see the world in color for the first time. This would open up an entirely new experience. Similarly, seeing the world with the eyes of a physicist for the first time can enrich your experience of the world. Some TIP students have this experience, and that is my wish for you.

In 2010 one of my students was very excited when she observed a "glory", a very rare optical phenomenon that I had described in class, but which few people ever see. She sent me the photo below.

Photo of the Glory by Shea Franklin, May 3, 2010. The Glory is best seen when one is looking down from a mountaintop onto one's shadow on clouds below. The back scattering of sunlight by the cloud creates a halo around the observer's head, but, because the effect is very directional, no halo is seen around the head of the person standing next to the observer. In this case, particles in the pavement functioned like water droplets in a cloud to produce the same effect.

Physics is applied in all of the sciences, especially the life sciences. That's why this is a required course. Understanding physics will give you a deeper understanding of topics in your major. I provide some applications of physics to physiology throughout the course, for example, forces exerted by muscles in the human body (ch. 10), the human circulatory system (ch. 11), and the human eye (ch. 26). Hopefully, these applications and the importance of physics as a foundation for your other science courses will help to motivate you.

A physics project on a topic of interest to you will, I hope, provide you with additional motivation. The project will also improve your understanding of physical principles by applying them.

Physics Project
A single group project is due during the last week of the semester. Groups of four correspond to tables in the classroom. The project presentation will be a five minute video uploaded to YouTube that will be shown to the class during the last week of class. The project can be most anything that is an application of the physics covered this semester. The physics of soccer, the physics of skateboarding, and the physics of dance are a few examples. In the second semester of physics, brainwave measurement, electric cars, and magnetic resonance imaging could be good projects. In addition to the video, each group will also submit a brief paper explaining the physics of their project. The topic you choose and a brief plan should be submitted on Oct. 11th. The main purpose of the project is for you to explore something that you are already interested in and to apply physics to attain a deeper understanding of it. My hope is that this will build your interest in physics.

Thinking Journal
Periodically you will be asked to make a thinking journal entry, reflecting on the development of your thinking in physics, the extent to which you are achieving course goals and your own personal goals, and answers to prompts that I will supply. Each entry will be a short Word document that you share with me, using the Dropbox website. (I will show you how to do this during the first lab.)

Games: MIND, Sudokus, Connecting the Dots
These games are introduced in the first week of class. Though they are not directly related to physics, they are useful in developing ways of thinking that are helpful in physics. Specifically, these games require attention to detail, holding several thoughts in working memory simultaneously, and planning strategies, rather than just trial and error. Attention to detail, combining information in working memory, and planning are essential elements of thinking in physics. Most students find the games challenging, yet enjoyable. The games are optional. You will be given an opportunity for a small amount of extra credit based on the games. We will devote very little class time to the games. Work on the games may be more important for some students than others.

Reading the Text; Online Reading Tests
In contrast to many other science courses, you will begin your study of each new topic not with a lecture, but by reading an assigned part of the text. This will require some effort on your part, but it will make class time much more productive. Your first reading of a chapter should focus on concepts. You can skim over the worked examples in the first reading. You don't need to know how to solve complex problems right away. Instead you should try to get a good understanding of the basic concepts that will provide the foundation for problem solving later. To test your understanding, you will answer several multiple-choice questions related to the material you read. These questions, which you will access on the university's Blackboard website, MYLMUConnect, are pretty basic, and if you have read the material carefully, you should be able to correctly answer all the questions. If you do not answer all the questions correctly on the first try, that's OK because you have two more opportunities to take the test. Before retaking the test, however, you should reread the parts of the text related to questions you are not sure about. It's also a good idea at this point to try some of the end of chapter questions (not problems), which cover more subtle points. These questions are of the same type as those discussed in the concepts classes.

Concepts Class
After each reading assignment, a class is devoted to developing a deeper understanding of the concepts covered in the reading, without detailed problem solving or computation. I will explain concepts, then pose multiple-choice concept questions designed to stimulate interest and promote discussion. You will use a personal response (clicker) system to submit your anonymous answer, and the distribution of answers will be displayed to the class as a bar graph. Typically

students find many of the questions challenging, and significant numbers choose wrong answers. After the initial polling, you will be asked to talk to your partner, arguing your case for your chosen answer. When you and your partner agree, you should discuss the question with others nearby who disagree. Students are polled again. Often many more students arrive at the correct answer on the second try. If that is the case, I will ask someone to explain their answer to the class. If significant numbers still get the wrong answer on the second try, I will provide a hint and ask you to reconsider the answer one last time. Eventually the class arrives at a correct answer through active engagement with the physics concepts embodied in the question. Why don't I just give you the answer? Because that doesn't work! The learning achieved by students in a Thinking in Physics class far exceeds that of students in a traditional science class where the answers are just given to them. Why is this? It's just the way the human brain works. It's a lot like a muscle in that respect. It has to do some heavy lifting to get stronger. More technically, when you learn you are building networks of neurons. An enormous amount of research over the past 20 years demonstrates that traditional teaching, a class in which the instructor only explains, just does not work.

After Concepts Class
After each concepts class, you should review the PowerPoint slides for that class, which are available on MyLMUConnect, and you should also review other conceptual end of chapter questions. At this point you should have a strong foundation for quantitative problem solving.

Homework Problems
Read the chapter again, this time reading the example problems very carefully. A great way to read the examples is to cover up the solution and try to create your own solution first. If you get stuck, which is likely, uncover just enough of the solution to get you started again. Working your way through the examples in that way will ensure that you are well prepared to tackle the end of chapter problems. Be sure that when you are doing homework problems you follow the Polya steps of problem solving (described in our second class meeting). Everyone wants to get an answer quickly, but slowing down and being very careful in the beginning of a problem will often make your problem solving more efficient and quite often even faster. Answers to most of the assigned problems are in the back of the book. If you are stuck, there are many resources for help. The solutions manual provides solutions to some of the assigned problems. These solutions should be used sparingly, in the same way I suggested you use the worked examples in the text.

Pencasts
I have recorded "pencast" solutions to selected problems. These are QuickTime movies on MyLMUConnect. The recorded audio and pen strokes can be started and stopped. A pencast provides all the elements of careful problems solving – formulating, planning, execution, and review. You should not merely view the movie. Instead, you will be asked to provide each step of problem solving before

viewing the recorded problem solving step. After viewing each step, you are asked to reflect on your solution, and whether your step included essential elements that were present in the pencast solution. For example, did your formulation include an adequate drawing? This element of metacognition, tied to a specific problem you have attempted, is helpful in developing a more systematic approach to problem solving.

Office Hours
You can see me for one-on-one help during office hours, or during lab sessions if you finish early, or by appointment. This can be especially helpful in diagnosing the kind of difficulty you are having with concepts or problems.

Problem Solving Class
I will guide you through group problem solving. We will work both assigned homework problems and some new, more challenging problems.

After Problem Solving Class
Following the problem solving class you should continue to work on problems and concepts, and prepare for a quiz the following week. You can utilize any of the resources: textbook examples, solutions manual, pencasts, and office hours.

Labs
TIP labs involve hands-on activities closely related to the physics covered in homework and in Concepts and Problem Solving classes. To better focus your attention on the underlying concepts, a short lab concept quiz is provided for each lab. No grade is attached to the quiz, which is taken at the end of the lab and then discussed by the class. The discussion is led by the TA or by me. Full credit is given for lab attendance, good faith effort, and participation in the review of the lab and concept quiz. Labs give you an opportunity to make sense of what you have been studying in a very concrete way. The quizzes help you to gauge your progress. Many labs will utilize computer simulations of experiments in addition to actual experiments. Excellent simulations on the PhET website can be reviewed after lab.

Quizzes, Tests, and Grading
The weekly quizzes encourage you to keep up with the material and make a serious effort on the homework. Because concepts and problems are both very important in physics, points on quizzes and tests will be approximately equally divided between concept questions and problems. In preparing for tests, you should review both concepts and problems. You should also include a review of lab write-ups and lab quizzes as part of your preparation.

Thinking in Physics tests are hard. Why? Because physics is hard, and any test that measures your understanding of physics must do so by probing for real understanding, not by simply asking you to regurgitate solutions to very specific problem types that you have memorized. Because Thinking in Physics tests are hard, scores on these tests tend to be much lower than in most other courses. 70% is a good score on these tests, indicating good understanding of physics concepts and good problem solving. 80% is an excellent test score. Although tests are harder and test scores are therefore lower, the distribution of grades in Thinking in Physics classes is similar to the grade distributions in other science classes. In recent Thinking in Physics classes the average grade has typically been B or B-.

You are not competing with others for a grade in this class. There is no limit to the number of A's that can be earned. If over half the class demonstrates an excellent understanding of physics, then over half the class will earn A's. By pushing your thinking, you will learn more, even though your test scores may not seem very high. Understanding this will require some adjustment in your thinking about the meaning of test scores.

My tests used to be open book. However, many students would spend the entire test thumbing through the book, looking for something to help them with a test problem, rather than just thinking about the problem. So now tests are closed book, but you can use a one page formula sheet.

You will have an opportunity to review your graded test, make corrections to anything that was wrong, and reflect on what it was about your thinking that led to mistakes on specific questions and problems. You can submit these corrections and reflections and I will give you 25% of any deducted test points for which you supply the correct solution and a reflection. In this way each test is not only an assessment of your understanding, but also an opportunity for further learning.

Anonymous Comments from Recent Students' Course Evaluations
"Loved the course."
"Hardest class I've ever taken."
"I had to work harder than in any other class this semester."
"Doing concept questions using clickers helped interest and learning of the material."
"I will definitely use the problem solving strategy outside of physics."
"Although I didn't like having weekly quizzes, they forced me to stay on top of the material and made studying for tests much more manageable."
"This course changed the way I perceive things around me. It made me pay attention to details of things I see happening or working everyday and made me grasp a bit of their nature. I enjoy the teaching method because it prompts my thinking."
"The conceptual thinking forced me to think out of the box."
"The way the class was taught really challenged me to understand more than plugging in numbers."
"Forced me to think from every angle."
"<u>Very</u> difficult but learned so so much. Loved learning about physics."

Advice to New Thinking in Physics Students from Former Students

"The most important thing a student can do to be successful in this class is to not get behind."

"Reading the text carefully, multiple times is very important. Also, when doing problems, don't just go through the motions, but think about why you do each step."

"Invest time into physics and go to office hours if a concept is not understood clearly."

"Don't be afraid to change your mindset; seeing things from a different perspective will really help you."

"Use Polya's method when attacking problems. Slow down and take your time during problems. It's not about finishing fast, but solving the problem correctly and understanding how you solved it."

"Spend a lot of time reviewing the material and when you feel like you've spent about the right amount of time, review some more. Use all resources available to you including the book and all it has to offer, the solutions manual, the pencasts and also your peers and the professor if need be."

"Start the project earlier in the semester so you don't rush to get it done last minute, read the chapters twice before taking the online quizzes."

"Understanding concepts is actually more important than doing problems because it shows the whole picture and will eventually help with problem solving."

"Don't stress out or worry about how 'difficult' the material is. If you do not do well on one test or one quiz it is not the end of the world. Everyone learns at their own pace and eventually it will click."

Appendix 3: PowerPoint Slides for the First Week

First Class

Welcome to PHYS 253 Please turn on your clickers **Music: Moonlight Sonata** 1st Movement (Wilhelm Kempff on YouTube) **Roll**	**A physics concepts diagnostic test** finding out what you already know
Our First Clicker Questions Do you have a strong interest in learning physics? A) yes B) no Do you have anxiety about taking this course? A) yes B) no	**Why Study Physics?**
Studying Physics can transform the way you see the world!	**Seeing in Black & White** (B & W photo of oil film) 
Seeing in Color (color version of the same photo) 	**What makes physics challenging?** It requires CHANGING: • your concepts about the physical world • the way you approach problem solving • perhaps the way you think about learning

A Physics Concept

Is a force required to keep an object moving? A) Yes   B) No

Turn on your hovercraft
& give it a push.

Physics Preconceptions

Everyone has some concepts related to force and motion that work well for everyday life, but which have limitations. How can we change these preconceptions?

Will your concepts change as a result of having the correct concepts explained to you?
A) always; B) usually; C) usually not.

Learning Physics

Can you learn physics by only paying close attention in class while everything is carefully explained to you?
A) Yes   B) No

Grading

If your overall average at the end of this course were 80%, what would be your grade in this class?   A   B   C

A roughly 80%
B roughly 70%
C roughly 60%
D roughly 50%

Questions?

Introduction to PHYS 253
This section is a special version:
Thinking in Physics

- Syllabus & Schedule
- A Learning Guide
- Text: Physics Fundamentals
- Sign up for an interview

Course Objectives;
Expected Learning Outcomes

- Understand physics concepts
- Apply physics principles in problem solving
- Appreciate applications of physics
- Develop scientific reasoning skills
- Develop general problem solving ability

Thinking in Physics students develop scientific reasoning skills

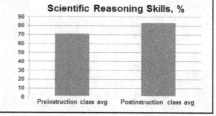

PowerPoint Slides for the First Week   213

Thinking in Physics students learn physics concepts 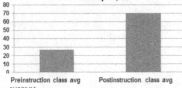	Results vary for individual TIP students 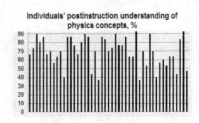
Thinking in Physics course structure For each new physics topic: • Reading Text • Concepts & Clicker Questions • Homework Problems • Problem Session • Lab • Quiz or Test	**Reading Test** Read assignment, aiming for conceptual understanding Reading test on MyLMUConnect End of chapter concept questions

Concepts Class
 	
Lecture, Clicker Questions, Discussion
 	
Powerpoints can be reviewed after class on MyLMUConnect
	Homework Problems
	
	• Try worked examples in text first
	• Use Polya method
	• Resources:
	    –solutions manual (MyLMUConnect)
	    –pencasts (MyLMUConnect)
	    –office hours
	    –problems class

Problems Class • Develop problem solving strategies • Group work on problems	**Lab** • Hands on experience • Emphasizes concepts • Ungraded lab concept quizzes to confirm understanding • Computer simulations complement experiments

Tests

- 50% conceptual questions, 50% problem solving
- Tests are challenging: 70% is good; 80% is excellent
- Tests are learning experiences
- Corrections + reflections can earn 25% of missed points. In reflections you engage in metacognition.

Metacognition: Self-awareness of one's thinking processes & the use of this awareness to regulate these processes

Success in Thinking in Physics requires metacognition

Other course elements

- Thinking Journals
- Group Project
- Games

Thinking Journal – reflect on the development of your thinking in physics

1) Tell me something about yourself.
2) Have you had a good high school physics course?
3) Your goals for yourself in this course.
4) How much time each week do you plan to study outside of class?
5) A concept or belief that you once thought was correct and no longer do, & how that came to change.
6) Your thoughts on the course objectives and learning guide.

Group Project

Working in groups of four, make a 5 minute video on the physics of …. (choose something you're interested in, for example: soccer, skateboarding, dance, forces exerted by muscles in the human body, …)

Games: next class

Student Comments about Thinking in Physics

"hardest class I've ever taken"
"This course changed the way I perceive things around me. … and made me grasp a bit of their nature."
"The conceptual thinking forced me to think out of the box."
really challenged me to understand more than plugging in numbers."
"Forced me to think from every angle."
"<u>Very</u> difficult but learned so so much."

Homework Tonight

Read Course Objectives, Schedule, & Learning guide

Send me your first Thinking Journal via Dropbox

Second Class

Music: Saint Saens Rondo Capriccioso Violin:Janine Jansen on YouTube	**Problem Solving** **Kinds of Problems** • Everyday problems • Brainteasers • Math problems • Physics problems • Research problems • Diagnosing health problems
When you first look at a problem on a physics test, should you be able to know how to solve it? A) Yes B) No	If you look at a "problem" and immediately know how to solve it, it's *not a problem*; for you it's an exercise. Not initially knowing how to solve a problem does not mean that you won't be able to solve it.
We will develop a general, helpful approach to problem solving?	**Getting There On Time** a practical problem Suppose you live off campus somewhere in the South Bay area. Your class schedule requires you to be on campus by 9 am. Each day you find yourself stuck in freeway traffic, and you are often late for class. What can you do?
A puzzle A bear starts at point P and moves 100 m directly south. The bear then moves 100 m east. Finally the bear moves 100 m north, arriving at exactly the same point P that he started at. What color is the bear?	How important to your success are your beliefs about problem solving? Beliefs about the problem? Beliefs about yourself? Be specific about what kinds of beliefs are important. Is it ever important to just leave a problem alone, to sleep on it?

Let's see if we can come up with a general approach to problem solving, some important general steps that we can take in trying to solve any problem – in math, in physics, or in life. Polya, a famous mathematician, came up with four such steps. Let's see if we come up with similar ones.

    I   II   III   IV

Polya's Method

I Formulate the Problem
   what we know and what we want to find, draw a sketch
II Plan a Solution
   without solving, explore ways that might work until you have a plan
III Execute the plan
IV Review the solution

Skill Building Games

Connect the Dots

Sudokus

MIND Challenge

Appendix 4: Syllabus and Schedule

First Semester Only (2nd semester syllabus/schedule on ThinkingInPhysics.com)

PHYS 253
Corequisite: MATH 122 or MA 131
Professor Vincent Coletta   vcoletta@lmu.edu   Seaver 105
Text: *Physics Fundamentals,* Coletta
Any changes that need to be made will be announced in class.

Course Objectives; Expected Learning Outcomes

This course is designed to give you an opportunity to: **1)** gain a clear **conceptual understanding** of physical principles, **2)** enable you to apply those principles in **solving physics problems**, **3)** develop an **appreciation for physics** in your everyday life, and for its application in other fields of science, **4)** develop your **scientific reasoning skills;** and **5)** improve your **general problem solving ability.**

Course Organization

The study of each new topic begins with an assigned reading from the text and an online reading test. Then in Concepts classes we will develop physics concepts, with class discussion questions utilizing a clicker response system. Following each concepts class is a problem oriented class and a lab on the same topic. Each student will have a partner. Partners will work together on classroom discussion questions and problem solving. You are encouraged to work together outside of class as well. But you should begin your study of each reading and problem assignment independently, so that you are not relying too much on your partner. Topics covered in the course include kinematics, relative motion, Newton's laws, force laws, momentum, energy, and torque.

Grading

Quizzes are open book. All other quizzes and tests are closed book, except that you are allowed to use a one page (8 1/2" x 11") formula sheet for midterm tests and two pages of formulas for the final exam. Grades will be based on the following (approximate) point system:

		Points
Quizzes		70
Reading tests		60
Project		25
Thinking journals		20
Clicker questions		20
Interview		5
Midterm Tests	2×100 points	200
Final Exam		200
Lab		100
	Total	700

Email Communication
It is your responsibility to read messages sent to you at your LMU Lion email address. If you use some other email account, be sure to have your Lion email forwarded.

Attendance
Perfect attendance is strongly recommended.

Honesty
Every effort will be made to promote academic honesty. Safeguards against cheating are part of the course design. Cheating in any form will result in a minimum penalty of failing the course.

Classroom Etiquette
Class begins promptly as scheduled. Please be on time. No use of cell phones or computers, except when part of class work.

Office Hours
Monday: 4:00 – 5:00, Tuesday: 3:00 – 4:00, Thursday: 1:30 – 2:30

PHYSICS 253 SCHEDULE
For each new topic the following sequence is followed: Read Text, Reading Test, Concepts Class, Reread Text and Examples, Homework Problems, Review Solutions and Pencasts, Problem Session, Additional Problems, Lab, Lab Concept Quiz, PhET, Review all, Quiz or Test

		Homework
1	8/27 Tues., Introduction to Thinking in Physics	1st Thinking Journal
	Course Objectives, Learning Outcomes	Schedule interview
		Read syllabus and study guide
2	8/29 Thurs. Introduction to problem solving	Read Introduction,
	Lab 1 Introduction to Lab	Meas. & Units, & Ch.1
		Reading Test 1
3	9/3 Tues. Concepts 1: Meas & Units, motion, vectors	Problems, Meas. & Units & Ch.1
4	9/5 Thurs. Problems: Meas & Units, Ch.1	Read Sects 2-1, 2-2,
	Lab 2 Variables, motion	2-4, 3-1, & 3-4
		Reading Test 2
5	9/10 Tues. Concepts 2: Acceleration, relative motion	Problems, 2-1,2-2,2-4 3-1, 3-4
6	9/12 Thurs. Quiz1 on ch 1; Probs: Accel & rel motion	Read Sects. 4-1 to 4-5
	Lab 3 Acceleration, relative motion	Reading Test
7	9/17 Tues. Concepts 3: Newton's laws of motion	Problems 4-1 to 4-5
8	9/19 Thurs. Quiz 2 acc & rel motion; Prob: Ch. 4, sect 1-5	Read Sects 4-6 to 4-8

Syllabus and Schedule

9 9/24 Tues. Concepts 4: Force laws, appl of N's laws   Problems, 4-6 to 4-8

10 9/26 Thurs. Quiz3,motion & force; Problems: 4-6 to 4-8   Read 2-3 & 3-2
       Lab 5 Newton's 2nd law   Reading Test 5
11 10/1 Tues. Concepts 5: Free fall, projectile motion   Problems, Sect 2-3, 3-2
12 10/3 Thurs. Problems, Free fall, projectile motion
       Lab 6 Free fall, projectile motion
13 10/8 Tues. **Test #1, Ch. 1-4**   physics project plan due 10/11
14 10/10 Thurs. Review Test; project plan due   Read Ch. 5, sect 1
       Lab 7 Projectiles   Reading Test 6
15 10/15 Tues. Concepts 6: Friction   Problems, Ch. 5, sect 1
16 10/17 Thurs. Problems, Ch. 5, Sect 1   Read sect 3-3, 5-2, & 5-3
       Lab 8 Friction   Reading Test 7
17 10/22 Tues. Concepts 7: Center of Mass & Circ Motion   Probs, 3-3,5-2, 5-3
18 10/24 Thurs.Quiz 4 Friction;Probs,Sect 5-2,5-3   Read sect 6-1,6-2,8-1,8-2
       Lab 9 Circular Motion   Reading Test 8
19 10/29 Tues. Concepts 8, Gravitation & Momentum   Problems Ch. 6 & 8
20 10/31 Thurs. Quiz 5 C of M & Circ motion; Prob, Ch. 6 & 8
       Lab 10 Rollerblade Platform
21 11/5 Tues. **Test 2 on Ch. 4, 5, 6, & 8**   Read Ch. 7, sect 1 & 2
       Reading Test 9
22 11/7 Thurs. Concepts 9: Energy, sect 1 & 2   Read Ch. 7, sect 4 – 7 & 12-3
       Lab11 Uncertainty   Reading Test 10
23 11/12 Tues. Concepts 10, Energy, sect 4 – 7   Problems, Ch. 7
24 11/14 Thurs.Problems, Ch. 7   Read sect 9-2 & Ch. 10
       Lab 12 Energy   Reading Test 11
25 11/19 Tues. Concepts 11, Torque & Static Equil   Problems, Ch. 9 & 10
26 11/21 Tues. Quiz 6. Energy; Probs, Ch. 9 & 10   Read Ch. 11, sect 1-3 & 4-7
       Reading Test 12
27 11/26 Thurs. Concepts 12, Fluids   Problems, Ch. 11
       Lab 13 Torque & Static Equilibrium
       Thurs., 11/28 THANKSGIVING HOLIDAY
28 12/3 Tues.Quiz 7 on Torque & Static Equil; Probs, Ch. 11

29 12/5 Thurs.Review; Post Assessment
       Lab 14 Present Projects
FINAL EXAM   **Tues, 12/10, 11:00 am**

HOMEWORK QUESTIONS
All odd-numbered end of chapter questions should be answered when that chapter is assigned for reading.

HOMEWORK PROBLEMS

Chapter	Assigned Problems	Solutions Manual	Pencasts
Measurement	7,9,11,13,15,17,19,23,27		
1	1,3,5,9,13,17,19,23,25,29, 31,33,35,37,39,45,49	3,17,29,35,37,45	5,17,34,53
2 (sect 1,2,& 4)	1,3,9,13,15,19,35,37	19,35	15,37
3 (sect 1 & 4)	1,3,27,31	27	5,31,33
4 (sect 1 to 5)	1,3,5,7,9,11,13,15,17	3,9,13,17	7,19
4 (sect 6 to 8)	20,21,23,25,29,31,35,46	23,25,31,35	24,31,40,42
2 (sect. 3)	21,25,27,29,55	21,25	27,33
3 (sect 2)	9,13,15,17,51	9,15,17,51	15,38
5 (sect 1)	1,3,5,7,9,39,41	3,9	7,10,40
3 (sect 3)	21,25,41	41	41
5 (sect 2 & 3)	21,23,27,29,31,36	23,27,31	24,35
6	1,5,11,17,21,23,25	1,5,11,23	24,36
8	1,3,5,9,11,13	5,11	10
7 (sect 1 & 2)	1,5,11,15,17,19,21	19	13,59
7 (sect 4 to 7)	31,33,37,39,43,45	31,39,43	37,57
9 (sect 2)	9,11,15	9,15	
10	1,3,9,15,19	3,9	8,16
11	1,3,5,7,9,13,19,21,23, 49,51,53,54	1,5,13,21,51	

Appendix 5: Seating Chart

Photos copied from the LMU website are copied into each of the oval spaces below, with students' names inserted beneath their photos.

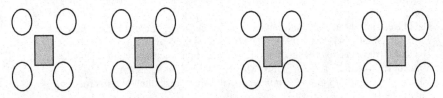

Please sit in your assigned seat and introduce yourself to the person sitting across from you.

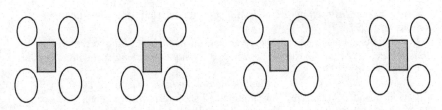

References

1. Richard P. Feynman, Robert B. Leighton, and Matthew Sands, *The Feynman Lectures on Physics* (Addison-Wesley, Reading, Mass., 1965).
2. Paul G. Hewitt, *Conceptual Physics* (Addison-Wesley, Reading, Mass., 1971).
3. Arnold B. Arons, *A Guide to Introductory Physics Teaching* (John Wiley & Sons, New York, 1990).
4. David Hestenes, Malcolm Wells, and Gregg Swackhamer, "Force Concept Inventory," *Phys. Teach.* **30**: 141-158 (1992). The 1995 revision by Halloun, Hake, Mosca, & Hestenes, is the version most commonly used and the one referred to in this book.
5. Vincent P. Coletta, *College Physics* (McGraw-Hill, Boston, 1995).
6. Vincent P. Coletta, *Physics Fundamentals* (Physics Curriculum and Instruction, Lakeville, MN, 2010).
7. Richard R. Hake, "Interactive-engagement vs. traditional methods: A six-thousand-student survey of mechanics test data for introductory physics courses, " *Am. J. Phys.* **66**: 64-74 (1998).
8. Eric Mazur, *Peer Instruction: A User's Manual* (Prentice Hall, Upper Saddle River, New Jersey, 1997).
9. B. Inhelder and J. Piaget, *The Growth of Logical Thinking from Childhood to Adolescence; An Eessay on the Construction of Formal Operational Structures.* (Basic Books, New York, 1958).
10. A. E. Lawson, *Classroom Test of Scientific Reasoning (revised edition).* 2000.
11. V. P. Coletta and J. A. Phillips, "Interpreting FCI scores: Normalized gain, preinstruction scores, and scientific reasoning ability," *Am. J. Phys.* **73**: 1172 – 1182, Dec. 2005.
12. V. P. Coletta, J. A. Phillips and J. J. Steinert, "Why you should measure your students' reasoning ability," *Phys. Teach.*, **45**: 626-629 (2007).
13. M. A. Dubson and S. J. Pollock. "Can the Lawson test predict student grades?" AAPT *Announcer* **36** (2): 90 (2006).
14. K. Diff and N. Tache, *AIP Conf Proc.* **951**: p. 85-87 (2007).
15. P. Pamela and J. Saul, "Interpreting FCI normalized gain," *AAPT Announ.* **36**: 89 (2006).
16. B. A. Pyper, "Changing scientific reasoning and conceptual understanding in college students,"*AIP Conf Proc.* **1413**: 63- 65 (2011).
17. P. Nieminem, A. Savinainen, and J. Viiri, "Relations between representational consistency, conceptual understanding of the force concept, and scientific reasoning," *Phys. Rev. ST: Phys. Educ.* **8**: 010123-1– 010123-10 (2012).
18. V. P. Coletta, J. A. Phillips and J. J. Steinert, "Interpreting force concept inventory scores: Normalized gain and SAT scores," *Phys. Rev. ST: Phys. Educ.* **3**: 010106-1– 010106-5 (2007).
19. B. Kurtz, dissertation, University of California (1976).
20. R. Karplus, "Science teaching and the development of reasoning," *J. Res. Sci. Teach.* **14** (2): 169-175 (1977).
21. R. Feuerstein, Y. Rand, M. B. Hoffman, and R. Miller, *Instrumental enrichment: An intervention program for cognitive modifiability* (University Park Press, Baltimore, 1980).
22. L. S. Vygotsky, *Mind in Society: The Development of Higher Psychological Process* (Harvard University Press, Cambridge, 1978)

References

23. P. S. Adey, and M. Shayer. *Really Raising Standards: Cognitive intervention and academic achievement* (Routledge, London, 1994).
24. Kun Yuan et al, "Working memory, fluid intelligence, and science Learning," *Ed. Res. Rev.* **1**: 83-98 (2006).
25. R. W. Engle and M. J. Kane, "Executive attention, working memory capacity, and a two-factor theory of cognitive control," *Psychology of Learning and Motivation*, **44**:145-199 (2003).
26. D. A. Bors and T. L. Stokes, "Raven's Advanced Progressive Matrices: Norms for first year university students and the development of a short Form," *Educ. & Psych. Meas.* **58**(3): 382-398 (1998).
27. S. M. Jaeggi, M. Bueschkuehl, J. Jonides, and P. Shah, "Short- and long-term benefits of cognitive training," *Proceedings of the National Academy of Sciences*, 108, 10081-10086. Doi:10.1073/pnas.11033228108 (2008).
28. C. L. Stephenson and D. F. Halpern, "Improved matrix reasoning is limited to trainng on tasks with a visiospatial component," *Intelligence* **41**: 341-357 (2013).
29. S. M. Jaeggi, M. Bueschkuehl, P. Shah, and J. Jonides, "The role of individual differences in cognitive training and transfer," *Mem. Cogn.*, DOI 10.3758/s13421-013-0364-z (2013).
30. C. S. Dweck, *Self-theories: Their Role in Motivation, Personality and Development* (Psychology Press, Philadelphia, 1999).
31. M. Buschkuehl, L. Hernandez-Garcia, S. M. Jaeggi, J. A. Bernard, and J. Jonides, "Neuraleffects of short-term training on working memory," *Cogn. Affect. Behav. Neurosci.* DOI 10.3758/s13415-013-0244-9 (2014).
32. V. P. Coletta, "Reducing the FCI Gender Gap," *AIP Conf. Proc.* (2013).
33. D. Hestenes and M. Wells, "A mechanics baseline test," *Phys Teach* **30**:159-196 (1992).
34. SAT data from The College Board web site (www.collegeboard.com), accessed 2013.
35. C. M. Steele and J. Aronson, "Stereotype threat and the intellectual test performance of African Americans,"*J. Pers. Soc. Psych.* **69**: 797-811 (1995).
36. S. E. Osborn Popp, D. E. Meltzer, and C. Megowan-Romanowicz, " Is the Force Concept Inventory biased? Investigating differential item functioning on a test of conceptual learning in physics,"presentation, Am. Ed. Res. Assoc. (2011).
37. V. P. Coletta, J. A. Phillips and J. J. Steinert, "FCI Normalized Gain, Scientific Reasoning Ability, Thinking in Physics, and Gender Effects,"*AIP Conf Proc.* **1413**: p. 23- 26 (2011).
38. A. Miyake et al, "Reducing the gender achievement gap in college science: A classroom study of values affirmation," *Science,* **330**: 1234 -1237 (2010).
39. L. E. Kost-Smith et al, "Gender Differences in Physics 1: The Impact of a Self-Affirmation Intervention,"*PERC Proc.* **1289**: 197- 200 (2010).
40. J. V. Mallow, "Gender-related science anxiety: A first binational study," *J. Sci. Ed. & Tech.* **3** (4): 227-238 (1994).
41. P. Haussler and L. Hoffmann, "An intervention study to enhance girls' interest, self-concept, and achievement in physics classes," *J. Res. Sci. Teach.*, **39**: 870-888 (2002).
42. G. Polya, *How to Solve It* (Princeton U. Press, Princeton, NJ, 1945).
43. www.mindresearch.net.
44. http://phet.colorado.edu.
45. To be published.
46. D. Clayson, "Student evaluations of teaching: Are they related to what students learn? A meta-analysis and review of the literature." *Journal of Marketing Educaation,* **31**(1): 16-29 (2009).

47. Private communication.
48. M. S. Sabella, "Implementing tutorials in introductory physics at an inner-city university in Chicago," *AIP Conf. Proc.* 79 (2002).
49. M. S. Sabella and G. L. Cochran, "Evidence of intuitive and formal knowledge in student responses: examples from the context of dynamics," *AIP Conf. Proc.* **720**: *79 – 83* (2003).
50. R. K. Thornton and D. R. Sokoloff, "Assessing student learning of Newton's laws: The force and motion conceptual evaluation and the evaluation of active learning laboratory and lecture curricula" *Am. J. Phys.* **66** (4): p. 338-352 (1998).
51. M. S. Sabella and A. G. Van Duzor, "Cultural toolkits in the urban physics learning community," *AIP Conf. Proc.* (2013).

Photo Credits

Chapter 4: Page 26, Jon Rou/Loyola Marymount University.

Chapter 5: Page 29, MIND Research Institute.

Chapter 10: Page 72, © 2010 MIT/Courtesy of MIT Museum. Page 86, Richard Megna/Fundamental Photographs. Page 88, Christopher Futcher/Getty Images.

Text Credits

Chapter 1: Page 2, Source: Carl Sagan. [Adapted from Pat Duffy Hutcheon's article in Humanist in Canada (Autumn 1997), p.6-9; 33]. Page 2, Source: Arnold B. Arons, A Guide to Introductory Physics Teaching (New York: John Wiley & Sons, 1990). Page 5, Source: Based on V. P. Coletta and J. A. Phillips, "Interpreting FCI Scores: Normalized Gain, Pre-instruction Scores, & Scientific Reasoning Ability.," Am. J. Phys.73: 1172 –1182, Dec. 2005. Page 6, Source: A. E. Lawson, Classroom Test of Scientific Reasoning (Revised Edition). 2000. Page 6, Source: V. P. Coletta and J. A. Phillips, "Interpreting FCI Scores: Normalized Gain, Pre-instruction Scores, & Scientific Reasoning Ability.," Am. J. Phys.73: 1172 –1182, Dec. 2005. Page 7, Source: V. P. Coletta, J. A. Phillips, and J. J. Steinert, Interpreting force concept inventory scores: Normalized gain and SAT scores, Phys. Rev. ST: Phys. Educ. 3: 010106-1– 010106-5 (2007). Page 9, Source: R. Feuerstein, Y. Rand, M. B. Hoffman, and R. Miller, Instrumental Enrichment: An Intervention Program for Cognitive Modifiability. (Baltimore: University Park Press, 1980). Page 10, Source: M. Buschkuehl, L. Hernandez-Garcia, S. M. Jaeggi, J. A. Bernard, and J. Jonides, "Neural effects of short-term training on working memory.," Cogn. Affect Behav. Neurosci. DOI 10.3758/s13415-013-0244-9 (2014).

Chapter 2: Page 11, Source: Richard Buckminster Fuller. [P.M. Senge, A. Kleiner, C. Roberts, and B. Smith (1994). The Fifth Discipline Fieldbook: Strategies and Tools for Building a Learning Organization. New York, NY: Doubleday Currency, p.28]. Page 12, Source: Based on V. P. Coletta, J. A. Phillips, and J. J. Steinert, Interpreting force concept inventory scores: Normalized gain and SAT scores, Phys. Rev. ST: Phys. Educ. 3: 010106-1– 010106-5 (2007). Page 13, Source: V. P. Coletta, "Reducing the Gender Gap.," AIP Conf. Proc. (2013). Page 15, Source: V. P. Coletta, "Reducing the Gender Gap.," AIP Conf. Proc.(2013). Page 16, Source: Richard R. Hake, "Interactive-engagement vs. traditional methods: A six-thousand-student survey of mechanics test data for introductory physics courses.," Am. J. Phys. 66: 64-74 (1998).

Chapter 3: Page 17, Source: Marie Curie [As quoted in Madame Curie : A Biography (1937) by Eve Curie Labouisse, p. 69]

Chapter 4: Page 26, Source: John Dewey [John Dewey, Democracy and Education, New York: Macmillan Company, 1944, p. 167.]

Chapter 5: Page 29, Source: Plutarch [Plutarch: Moralia, Volume II, (Loeb Classical Library, 1928)]

Chapter 6: Page 37, Source: Lloyd Alexander [Lloyd Alexander, The iron ring (New York, N.Y. : Dutton Children's Books, ©1997)]. Page 46, Source: Dennis E. Clayson ["Student Evaluations of Teaching: Are They Related to What Students Learn? A Meta-Analysis and Review of the Literature," Journal of Marketing Education 31, no. 1: 16-30]. Page 46, Source: Carl Wieman.

Chapter 7: Page 47, Source: Desmond Tutu [Tutu, Desmond (1999). No Future Without Forgiveness. Image. ISBN 0-385-49690-7]

Chapter 8: Page 52, Source: Albert Camus [Albert Camus, The myth of Sisyphus and other essays (New York : Random House, 1942)]

Chapter 9: Page 56, Source: Fran Lebowitz [Fran Lebowitz, The Fran Lebowitz reader (New York: Vintage Books, 1994.)]

Chapter 10: Page 71, Source: Alan Alda [This quote was said by Alan Alda in his speech he gave at his oldest daughter's college commencement at 'Connecticut College' in 1980.]

Chapter 11: Page 118, Source: Albert Einstein [Albert Einstein, Léopold Infeld, The evolution of physics: the growth of ideas from early concepts to relativity and quanta (Simon and Schuster, 1938), p.92.]

Chapter 12: Page 122, Source: Jessamyn West [Jessamyn West, The Quaker Reader, (The Viking Press, New York, USA, 1962), p.2]

Chapter 13: Page 185, Source: Quote by Art Williams, formed NBA basketball player.